BOARD OF EDUCATION

C000258031

HANDBOOK OF THE COLLECTIONS ILLUSTRATING

AGRICULTURAL
IMPLEMENTS & MACHINERY

A BRIEF SURVEY OF THE MACHINES AND IMPLEMENTS
WHICH ARE AVAILABLE TO THE FARMER
WITH NOTES ON THEIR DEVELOPMENT

BY
A. J. SPENCER, M.I.Mech.E.
AND
J. B. PASSMORE, M.Sc.
(*University of Reading*)

Agricultural Tools and Machinery

Farming has an incredibly long history. Beginning around 3000 BC, nomadic pastoralism, with societies focused on the care of livestock for subsistence, appeared independently in several areas in Europe and Asia. This form of farming utilised basic implements, but with the rise of arable farming, agricultural tools became more intricate. Between 2500 and 2000 BC, the simplest form of the plough, called the ard, spread throughout Europe, replacing the hoe (simply meaning a 'digging stick'). Whilst this may not seem like a revolutionary change in itself, the implications of such developments were incredibly far reaching. This change in equipment significantly increased cultivation ability, and affected the demand for land, as well as ideas about property, inheritance and family rights.

Tools such as hoes were light and transportable; a substantial benefit for nomadic societies who moved on once the soil's nutrients were depleted. However, as the continuous cultivating of smaller pieces of land became a sustaining practice throughout the world, ploughs were much more efficient than digging sticks. As humanity became more stationary, empires such as the New Kingdom of Egypt and the Ancient Romans arose, dependent upon agriculture to feed their growing populations. As a result of intensified agricultural practice, implements continued to improve, allowing the expansion of available crop varieties, including a wide range of fruits, vegetables, oil crops, spices and other products. China was also an important centre for agricultural technology development during this period. During the Zhou dynasty (1666–221 BC), the first canals were built, and irrigation was used extensively. The later Three Kingdoms and Northern and Southern dynasties (221–581 AD) brought the first biological pest control, extensive writings on agricultural topics and technological innovations such as steel and the wheelbarrow.

By 900 AD in Europe, developments in iron smelting allowed for increased production, leading to improved ploughs, hand tools and horse shoes. The plough was significantly enhanced, developing into the mouldboard plough, capable of turning over the heavy, wet soils of northern Europe. This led to the clearing of forests in that area and a significant increase in agricultural production, which in turn led to an increase in population. At the same time, farmers in Europe moved from a two field crop rotation to a three field crop rotation in which one field of three was left fallow every year. This resulted in increased productivity and nutrition, as the change in rotations led to different crops being planted, including vegetables such as peas, lentils and beans. Inventions such as improved horse harnesses and the whippletree (a mechanism to distribute force evenly through linkages) also changed methods of cultivation. The prime modes of power were animals; horses or oxen, and the elements; watermills had been initially developed by the Romans, but were significantly improved throughout the Middle Ages, alongside windmills – used to grind grains into flour, cut wood and process flax and wool, among other uses.

With the coming of the Industrial Revolution and the development of more complicated machines, farming methods took a great leap forward. Instead of harvesting grain by hand with a sharp blade, wheeled machines cut a continuous swath. And instead of threshing the grain by beating it with sticks, threshing machines separated the seeds from the heads and stalks. Perhaps one of the most important developments of this era was the appearance of the tractor; first used in the late nineteenth century. Power for agricultural machinery could now come from steam, as opposed to animals, and with the invention of steam power came the portable engine, and later the traction engine; a multipurpose, mobile energy source that was the ground-crawling cousin to the steam locomotive. Agricultural steam engines took over the heavy pulling work of horses, and were also equipped with a pulley that could power stationary machines via the use of a long belt. They did operate at an incredibly slow speed however, leading farmers to amusingly comment that tractors had two speeds: 'slow, and damn slow.'

From this point onwards, it has been the methods of powering machines, rather than the agricultural machines themselves, which have been the biggest breakthroughs in farming practice. The internal combustion engine; first the petrol engine and later diesel engines, became the main source of power for the next generation of tractors. These engines also contributed to the development of the self-propelled, combined harvester and thresher, or 'combine harvester' (also shortened to 'combine'). Instead of cutting the grain stalks and transporting them to a stationary threshing machine, these combines cut, threshed, and separated the grain while moving continuously through the field. Combines might have taken the harvesting job away from tractors, but tractors still do the majority of work on a modern farm. They are used to pull various implements – machines that till the ground, plant seed and perform other tasks. Besides the tractor, other vehicles have been adapted for use in farming, including trucks, airplanes and helicopters, for example to transport crops and equipment, aerial spraying and livestock herd management.

The basic technology of agricultural machines has changed little in the last century. Though modern harvesters and planters may do a better job or be slightly tweaked from their predecessors, todays combine harvests still cut, thresh, and separate grain in essentially the same way. However, technology is changing the way that humans operate the machines, as computer monitoring systems, GPS locators, and self-steer programs allow the most advanced tractors and implements to be more precise and less wasteful in the use of fuel, seed, or fertilizer. In the foreseeable future, there may be mass production of driverless tractors, and new advances in nanotechnology and genetic engineering are being used in the same way as machines, to perform agricultural tasks in unusual new ways. Agriculture may be one of the oldest professions, but the development and use of machinery has made the job title of *farmer* a rarity. Instead of every person having to work to provide food for themselves, in America for example, less than two percent of the population works in agriculture. But today, a single farmer can produce cereal to feed over one thousand people. With continuing advances in agricultural machinery, the role of the farmer continues on.

PREFACE

THE National Collections in the Science Museum are designed to illustrate the development of physical science and to show the ways in which this has been applied to various branches of industry. In each group the more important stages of development are represented by selected objects, while others taken from the practice of to-day complete its history. In this way the historical series is permanent in character, but the objects in the current series are continually being replaced from time to time by later examples.

Whenever possible, exhibits are arranged so as to be set in motion by visitors, and many of them are sectioned and opened up so that the inner construction and working parts can be studied.

The creation of a Museum of Science was proposed by the Prince Consort after the Great Exhibition in 1851, and in 1857 collections illustrating foods, animal products, examples of structures and building materials, and educational apparatus, were exhibited at South Kensington. Since then they have been continuously added to, especially in 1883, by the collection of machinery formed by the Commissioners of Patents, in 1900 by the Maudslay Collection of machine tools and marine engine models, and in 1903 by the Bennet Woodcroft Collection of engine models and portraits.

Agricultural tools go back to the earliest human civilizations, and modern agricultural machinery for the most part has been developed directly from these. The attempt has therefore been made to represent both the primitive tool and the modern machinery with so many of the important intermediate types as can be accommodated in the space that is available ; and although it is far from being complete the collection already contains many objects of great historical interest.

3

CONTENTS

LIST OF ILLUSTRATIONS

INTRODUCTION

AGRICULTURE is the oldest industry in the world, and has probably been the slowest to adopt new methods and new apparatus. Hand labour is slow and arduous, and is often wasteful both of labour and of material. Improvement has been effected by the introduction of many types of labour-saving machines, but the agricultural worker in England, and in many other countries, has still not reaped the full benefit of modern invention and is often unacquainted with the latest devices in farm machinery. The various processes in farming in their order of operation are : tillage operations consisting of ploughing, cultivating, harrowing, and rolling, which produce the requisite fineness of division and freedom from weeds of the soil. These are followed by manuring and seed-sowing, and later by reaping, threshing, and milling. In practically all these processes, hand labour has been replaced by machinery which is constantly being improved in engineering design and construction. The following notes have been written with the object of setting forth the development of some of the more familiar implements in use in this country.

Passing over the digging-stick and its development into the spade, the plough may be regarded as the oldest agricultural machine. Its history goes back to the earliest civilizations, and the first recorded plough is found on Egyptian monuments. It was found in this country in Saxon times complete with mould-board, coulter and wheels. The harrow appeared in its earliest form prior to the eleventh century, and neither of these implements shows any marked change for a considerable period. The eighteenth century was marked by the introduction of the corn-drill, the horse-hoe, and the thresher. In the early part of the nineteenth century a new era opened for agriculture following the general application of machinery of the late eighteenth century, when so many inventions in all industries resulted from the application of steam as a motive power. Throughout the nineteenth century many new types of agricultural machines were introduced and steam power began to replace the older forms of wind and water power. By the end of the nineteenth century the design of most agricultural machines was stabilized, so that the twentieth century is marked mainly by improvement in detail. In recent years, however, the internal combustion engine has been adapted for farm work, and the motor tractor has been developed, while the introduction of sugar-beet has necessitated the development of special implements for dealing with this crop.

PLATE I

By courtesy of the Editor of "Antiquity" and Dr. E. Cecil Curwen, M.A.

Skye Crofter using the Caschrom (Cat. No. 2).

HISTORICAL REVIEW

Ploughs.—The early history of the plough in Great Britain is obscure. It is comparatively easy to discover what kind of plough was used in Saxon times, and the form of the Roman plough is known, but the kind of implement used by the Briton both during and previous to the Roman occupation is still a mystery. It may be that the Roman occupation produced some change in the native plough, but whatever happened in those early days, its bearing on the development of the modern English plough is probably negligible. The Saxon introduced his own plough, and the evidence available suggests that it is from that implement, and from nothing which preceded it, that our present-day plough developed.

The earliest known reference to the Saxon plough is in the form of a riddle, said to be of eighth century origin and found in an old manuscript in the library of Exeter Cathedral. There are also illustrations in Saxon manuscripts of the period just previous to the Norman conquest, and the Bayeux Tapestry, probably eleventh century, shows a plough and a harrow (*see* cover). This evidence indicates that for some time previous to 1066 A.D. the plough of this country had a long share-beam (the part of the plough which slides on the ground), a sheath bracing the fore-end of the share-beam up to the main beam, a mould-board to turn the soil to one side, a second handle to give better control and perhaps to assist in holding the mould-board out to its work, and a pair of wheels on the axle of which the fore-end of the beam rested.

Fitzherbert, in his " Boke of Husbandrie," 1523 (*c.*), draws attention to the ploughs in use at that time, and illustrates a two-wheeled plough drawn by oxen.

There was little development, however, of the plough up to the seventeenth century, when Walter Blith, an officer in Cromwell's army, in addition to describing the old English plough, a massive implement needing six or eight oxen to draw it, drew attention to light two-horse ploughs in Norfolk and Suffolk (Nos. 6–7), which had probably come from the Netherlands in the sixteenth century. Such a plough was in existence in that country in 1590 A.D. and for some time previously there had been close communication between the Netherlands and England.

The appearance of these light ploughs in East Anglia was followed in the year 1730 by the patenting of the Rotherham plough by Disney Stanyforth and Joseph Foljambe (No. 8), which shows a great advance in the improved form of its frame and had a profound effect on plough design in this country. It was adopted in the West Riding of Yorkshire and was able to compete successfully with Norfolk and Suffolk ploughs in their own respective districts. It also stimulated the imaginations of John Arbuthnot, in England, James Small, in Scotland, and others, who set to work on the elucidation of the first principles of plough design, and performed an important work in leaving records and plans from which ploughs could subsequently be built ; they were, in a large measure, responsible for the break-away from the old heavy plough to the lighter type of implement.

The closing years of the eighteenth and the first half of the nine-teenth century was a period of very rapid development. In the year 1800 Plenty, of Newbury, designed the iron frame ; in 1803 Ransome, of Norwich (later of Ipswich), introduced the chilled plough-share (No. 10), and by 1825 the complete iron plough was firmly established. During this period there developed a differentiation between autumn and spring ploughing, with the production of a distinct type of plough for each process. A long-breasted plough (No. 16), with a long gently twisting mould-board, was designed to lay the slices on edge and unbroken in the autumn, whilst a plough with a bluff elliptical breast (No. 17), a short mould-board twisting sharply and sometimes armed with knives, was used to burst up the soil in the spring and to leave it as well pulverized as possible.

From that period to the present-day progress has taken the form of improvements in detail, the increased use of chilled cast iron and steel for various parts, and adaptation to new methods of traction.

Paring Ploughs.—The burying of surface growth, grass or weeds, has always been a difficult problem for the farmer. Prior to the eighteenth century the breast plough (No. 4) was used for this purpose, and, in fact, may be seen in use at the present time in the West of England. The breast plough is a hand implement, consequently the process is slow and expensive. The top three or four inches of soil with its growth of weeds or grass was skimmed off and either dried and burnt, or it was allowed to lie on the surface ; then when the ground was ploughed in the ordinary way, anything lying on the surface was raked into the furrow and buried.

Jethro Tull, in the early half of the eighteenth century, tried to overcome the difficulty with a four-coultered plough. The three extra coulters were fastened to the beam parallel to the original coulter and spaced uniformly across the furrow. They were set to cut not quite to the full depth, so that the inverted slice consisted of ribbons of turf buried by a slice of soil. The harrows which followed the plough broke down the soil to a more or less uniform layer, so completing the burying and killing of the vegetable matter. This four-coultered plough had a beam ten feet long, and in addition to its extreme clumsi-ness, there was the difficulty of maintaining a sufficiently sharp edge on its wrought-iron coulters. As a consequence it soon dropped out of use and was never revived. Paring ploughs were known in the seventeenth century, and in effect, substituted animal power for hand labour in the work of the breast plough. It is fairly certain that the broad share used for paring was imported from the Netherlands in

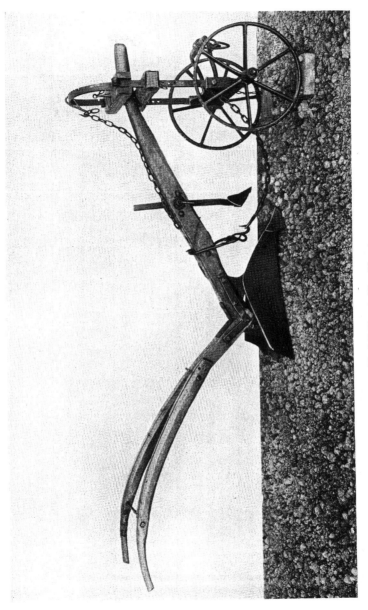

Norfolk Plough (Cat. No. 7).

Rotherham Plough (Cat. No. 8).

Small's Wooden Plough (Cat. No. 25).

the sixteenth century, and it is probable that this was the stimulus which led to the use of the fin on the Saxon spear share and paved the way for the later broad, flat common share.

One-way Ploughs.—The one-way plough is so built that it can be made to turn a slice either to the right or left. This facility makes it particularly adapted to sloping fields, although its use is not confined, and its application need not be limited, to hilly districts. The common plough is built to turn its furrow-slice to the right alone—a practice which probably began with the use of the mould-board. The reason for turning the slice to that side and not to the left is no doubt bound up with the fact that the driver of the team walks with them at his right hand, *i.e.* on their left.

The oldest type of one-way plough used in this country is the Kent plough (No. 21). It is mentioned in the sixteenth century and by many later writers, all of whom are either scornful or humorous about its clumsiness.

Double-ended ploughs (Nos. 22 and 23) probably originated in South Devon, and one of the specimens shown in the collection is a copy of one found in that district. Henry Lowcock of Westerland, Devon, is credited with having invented this type, but the only available evidence shows that he patented certain improvements to double-ended ploughs. As far as is known the plough was introduced into, or invented in, South Devon late in the eighteenth century ; it was certainly in existence in the year 1810, and a form of it, made in iron, was seen in use in the same district in 1923. A few are still in existence, if not in use, but they have been largely replaced by horse-drawn balance ploughs.

The balance plough, the turn-over plough, and one or two other types of one-way ploughs which have since practically died out, came into use during the nineteenth century. The first two types have survived but their use is largely confined to the south-western counties, although the principle of the balance plough has been used in the development of the steam plough. The Oliver type of one-way plough which on account of its lightness is suitable for garden work, is an American invention which was introduced into this country late in the nineteenth century.

Steam and Tractor Ploughs.—The horse-drawn two-furrow plough (Nos. 18 and 19) made its appearance in the late seventeenth century. Early in the following century attempts at steam ploughing began, and with the increase of power the demand arose for a multiple plough. Many devices were tried, but the standard form of steam plough is the balance type which is illustrated in the collections (No. 20). The tractor in the twentieth century was first used to draw two, three or even four-furrow horse ploughs usually controlled by a ploughman, but very soon it was found to be advantageous to adapt the plough to the higher speed of the tractor, to provide it with an automatic device for lifting it out of its work and dropping it in again at the head-land (the strip of unploughed land at the end of a furrow), and to bring the controls forward so that they could be operated by the tractor-driver, thus making it possible for one man to manage the tractor and the plough.

Other Implements of Tillage.—In the process of tilling the soil the farmer works in alliance with the forces of nature. He uses the plough to loosen the top-soil to a depth of about seven inches, to bury vegetable matter or manure so that it can rot and become, through the action of bacteria, available as plant food, and to expose fresh layers of soil at the surface to the action of the weather. These natural forces, by means of reactions which are partly chemical and partly physical, prepare the surface soil for the succeeding steps in the production of a seed bed. Certain substances in the soil are rendered more available as plant food for the crop which is to be sown, and the soil itself becomes more friable so that it can be broken down more readily into a fine state of division suitable for the reception of the seed. In a finely divided soil, in which the surface is loose and the lower layers are firmly packed, there is less possibility of individual seeds falling through crevices and being lost in the lower layers of the soil and the young seedling has a better chance of life in its earlier stages of growth. The implements which the farmer brings to the task are : the cultivator, the harrow, the roller and the horse-hoe. The cultivator does the heaviest part of the work of breaking down the soil and it is sometimes used without the preliminary loosening and inversion of the soil by the plough. Heavy harrows and medium harrows carry the work of pulverization further and also tend to compress the lower layers of the soil. The roller compresses the soil, allowing it to suck up moisture from the sub-soil and also assists in crushing clods. Light harrows and seed harrows loosen the surface, thus preventing evaporation and consequent loss of moisture from the soil, and they are also used to stir the surface to cover seed which has been sown. The horse-hoe is used for preserving the looseness of the surface layers and for uprooting weeds, and operates between the rows of the growing plants.

It would appear that as early as the tenth century a primitive harrow was in use in this country (No. 49). This consisted of a rectangular wooden frame, through the four sides of which spikes were driven, so that when the implement was dragged from a point near one corner the protruding spikes stirred the soil. There was also the bush harrow which consisted of thorn bushes bound to a wooden frame. Those two types of harrow were the only implements of their kind available for many centuries.

It is difficult to draw a hard and fast line between cultivators and harrows, but in a general way it may be said that the former does the rough work of breaking the soil after the plough, while the more superficial work of pulverizing and stirring is done by the latter. Cultivators are in general more heavily built and are provided with wheels, but harrows, except in the case of modern tractor harrows, are not usually provided with wheels.

12

An early attempt to make an implement on the lines of the modern cultivator was that of Edward Nicholas, a farmer, who in 1817 attached five small plough bodies by " spindles " to five beams, braced the beams together and mounted them on three wheels, one on each side behind, and one in the centre in front, these wheels being adjustable for depth. It is curious that the implement was not intended for breaking up the soil after the plough, but was designed for covering with " mould, wheat and other grain when sown." Neither this machine nor one patented by Dyson in 1818 seem to have made much impression, but apparently the triangular frame and the three wheels which are characteristic of the " Grubber," or more primitive type of present-day cultivator, were fixed at this period.

In 1830 numerous improved grubbers had begun to appear ; one made by Kirkwood (No. 47) had the three-wheeled, triangular form and was capable of being lifted out of work while it was in motion. This implement is interesting in that it shows two other progressive features. The frame could, when it was raised or lowered, be maintained in the horizontal plane, and the tines were slanted well forward. The importance of these two points lies in the fact that in the first case it was possible to adjust the depth of all tines equally, and in the second case the penetrative power of the implement no longer depended entirely on its weight.

Modern cultivators fall into four classes :—

1. The Grubber which is most closely allied to the early form ;
2. The Bar type, in which the triangular frame is attached by bell crank axles to the hind wheels, and two rows of tines are attached to the bar forming the base of the triangle ;
3. The horse-rake type, where the tines are attached to a dead axle between the two hind wheels and are lifted or lowered by rotation on this axle ;
4. The Canadian type which is not very common in this country.

Further improvements in the effectiveness of cultivator tines in pulverizing the soil, and separating out weeds, have been introduced in the form of the spring-tine and the spring-mounted tine. In the one case the tine itself is made of steel, and in the form of a flat ribbon, and in the second case a rigid tine is mounted against a strong spring. As it passes through the soil, the tine vibrates, and so sifts the soil more effectively, tending to shatter clods that are sufficiently brittle and to shake weeds free of soil and bring them to the surface.

The modern zig-zag harrow was apparently another product of the mid-nineteenth century. Earlier in the century the making of iron frames instead of wood had commenced. There was, however, always a certain difficulty with the rectangular form. It could only be drawn from one point, and was therefore liable to swing from side to side as it travelled. Triangular-shaped and diamond-shaped frames were attempted until the final zig-zag form was reached. Then a set of harrows was made up of two or three leaves, each leaf being attached to a single bar at two points.

13

As a rule modern harrows are made in grades ranging from heavy harrows for use after the plough or cultivator, to light seed harrows for covering seed after the drill. Light wooden-framed harrows are also in use to some extent at the present day. Heavy harrows are often provided with " duck-footed teeth," *i.e.* the end of the tooth is bent forward and splayed out a little to give it better penetration and grip of the soil, but for the most part harrow teeth are about six inches long, pointed, and straight. Chain harrows came into use in the early part of the nineteenth century, to replace the primitive bush harrow. In 1830 a chain harrow was being made which was fitted with numerous small corrugated discs, but this attempt to adapt the chain harrow for use on pasture land has been superseded by a system in which a number of short teeth are allowed to project from the horizontal plane of the chain harrow, to serve the purpose of tearing the surface of grass land.

The functions of a roller are to compress the soil, to crush clods which are brought to the surface by harrows or cultivators, and to press clods down into the soil so that they can absorb moisture from the surrounding soil, which enables them to fall to pieces under the action of the harrow. The power of compression which a roller has will depend on its weight, which, since its width is limited, will depend, in a solid stone roller, on its diameter. For clod-breaking, however, a small-diameter roller is more effective. The old wooden roller was never a very satisfactory implement, but in districts where suitable stone was abundant the more useful stone roller was common. The increasing use of cast iron early in the nineteenth century caused the introduction of cast iron cylinders, the roller being made in two halves. Even a roller divided in this way still had a tendency to rub the soil in turning at the headland and very soon rollers were built up of a number of wheels with flat and comparatively narrow rims mounted closely together on a common axle.

On the heavier soils the necessity for breaking clods outweighed the necessity for compression, and mention is made of a spiky roller in the eighteenth century. The introduction of the segmented iron roller led to the corrugation of the rims of the segments and the construction of the Cambridge roller (No. 58), and the Crosskill roller (No. 57). The latter was patented by Crosskill in 1841, and consists of a number of wheels with serrated rims which are mounted on a common axle alternately with wheels with plain rims. The axle holes of the former are of greater diameter than the axle while those of the latter fit the axle closely. In this way, as the roller turns, the spikes have a certain amount of radial play between the smooth-rimmed wheels thus preventing the surface of the roller from becoming clogged with soil.

The *Disc Harrow* is an implement which is used fairly widely in this country for pulverizing the surface and consolidating the lower layers of the soil. It consists of a number of saucer-shaped steel discs mounted on a common axle, and set at an angle to the line of

14

Kent Plough (Cat. No. 21).

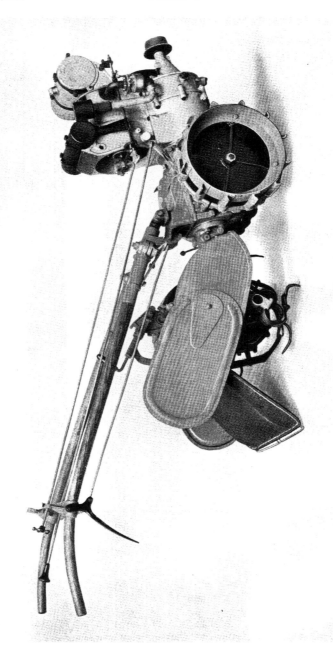

Simar Roto-tiller (Cat. No. 60).

travel. Two sets of discs, on separate axles, mounted on either side of the line of draught give sufficient resistance on most soils to comprise a three-horse implement; a tractor disc harrow is usually made up of four sets of discs. A disc roller, probably a forerunner of this implement, was in use in Cornwall in 1830 and consisted of thirteen flat iron discs mounted on a wooden roller, the diameter of which was about one half of that of the discs. It seems to have been used for pulverizing damp land.

Disc Ploughs consist of two or three saucer-shaped steel discs, larger than those of the harrow, but set at an angle in a similar manner. Such ploughs are useful for tearing up virgin soils or where fibrous roots or stones would make it impracticable to use the ordinary plough. They are not much used in this country.

Furrow Pressers.—Where a clover lea is ploughed up in the autumn preparatory to the sowing of wheat, it is necessary to press down the furrow slices so that there shall be no hollows in the lower layers of the soil. Robert Berriman, a wheelwright, in 1806, patented an implement for carrying out this operation, and the form of the modern furrow-press practically coincides with Berriman's specification. The implement consists of two heavy cast iron wheels with rims which are triangular in radial section, and are mounted on an axle with a third plain-rimmed wheel. It follows behind two ploughs, and is drawn by one horse. The plain wheel runs in the furrows while the two heavy wheels run between the crests of the previous three furrow-slices. This implement is also made with three presser wheels to take three furrows at a time.

Rotary Tillage.—Rotary tillage is a method of cultivation whereby the soil is reduced to a finely divided state in one operation. A machine which is used for the purpose is the Simar Roto-tiller (Nos. 60 and 61).

The orthodox method of preparing the soil for the sowing of the crops is a long and expensive process. The soil is ploughed in the autumn; throughout the winter, frost, rain, the drying action of wind, together with chemical reaction help to prepare the soil, though frequently excess of one or more of these agents causes hindrance. The soil may be ploughed again in the spring, it is torn up by the cultivator two or three times and the harrows and rollers are used alternately. Finally the required state of tilth is produced. In this way the same patch of soil may be treated at least eight times and with several different implements. The time occupied in the process will vary considerably, but estimated roughly for one acre it might be as much as 26 hours, occupying one man and two or three horses for three to four working days.

Rotary tillage has attracted the attention of many inventors in past years, but not until the advent of the internal combustion engine was it possible to apply it successfully.

The Simar roto-tiller has a working width of 20 inches, it moves at ¾ of a mile per hour, and is stated to pulverize the soil completely to any depth required from 2 to 10 inches, producing a tilth in one operation. It is claimed that in one working day, one man using one machine, can prepare an acre of soil for the sowing of seed, and the requisite soil and weather conditions have only to be obtained once.

This broad statement is merely one aspect of the problem, though probably apart from the economic side of the question, it is the most important. Rotary tillage was introduced into this country as a working proposition just after the War of 1914–18, and the present position is that the principles involved are being studied carefully.

SEED DRILLS AND MANURE
DISTRIBUTORS

Seed Sowing Machines.—Seed may be sown in one of three ways ; it may be broadcast, dibbled or drilled. Broadcasting is the scattering of the seed by hand or by a machine as uniformly as possible over the area sown ; wheat, and in fact all grain, has been sown in this way by hand from the very earliest times up to the present day. Larger seed, such as beans and potatoes, are sometimes dibbled ; holes are made in the soil at definite distances apart in rows, and a single seed potato or two or three beans are dropped into each hole. Drilling is a kind of compromise between broadcasting and dibbling ; the seed is sown in rows or drills, but is not spaced along the drill. When a crop is dibbled or drilled not only is there a considerable saving in the amount of seed used, but it is also possible to carry on cultivation of the soil between the rows of the growing crop.

There is evidence that from 1700–1200 B.C. the Babylonians used a drill plough,* but interest in drilling in this country was practically non-existent until it was stimulated primarily by the introduction of the turnip in the seventeenth century. Even then it was not until the end of that century that Jethro Tull, a cultivator of Berkshire, laid the foundations of " Drill-Husbandry."

Prior to Tull, Sir Hugh Plat, in 1600, had experimented, and recommended in his book, " The Setting of Corne," a method of dibbling wheat. In explaining the origin of dibbling wheat he states that a " silly wench," employed sowing carrots and radishes, accidentally dropped grains of wheat into some of the holes. These isolated wheat plants showed exceptionally luxuriant growth, and set Plat thinking so that he devised his method of dibbling wheat in the field. John Worlidge, in 1669, suggested the drilling of corn and outlined a machine for the purpose, but it is doubtful whether the machine was ever constructed.

The drill which Tull invented was described by him fairly fully, though it is no easy matter to build a drill on his not too exact specifications. His methods, and his drills seem to have lain unrecognised for some time after his death in 1764, until they were revived by certain Scottish farmers.

The last few years of the eighteenth century saw some quite considerable activity in this respect, for between 1780 and 1790 thirteen patents were obtained for seed-sowing machines. In the years between 1790 and 1800 no patents were granted, but interest seemed to revive again in the early years of the nineteenth century. The

* Publications of the Babylonian Section, vol. ii, p. 66, Pennsylvania University.

first patent was obtained by T. Proude in 1781 to cover a drill plough, next came J. Cooke in 1783 with a combined seed and manure drill, then in 1784 Horn patented a broadcast sower, and Wright a primitive kind of brush drill. By 1830 there were in use in this country the following types of seed-sowing machines :—

1. A device (No. 66) consisting of a hopper, seed delivery mechanism and actuating wheel which ran on the land, to be attached to a plough for sowing beans.

2. A bean-drill barrow (No. 65) which was pushed by hand up the furrow after the plough, and which could also be used for sowing turnips on the ridge.

3. A Northumberland turnip drill which sowed two rows at a time, was drawn by one horse, operated and driven by one man, and carried concave rollers for sowing on the ridge.

4. A multiple corn drill which on most soils was drawn by two horses and operated and driven by one man. The feed mechanism was rather primitive, but it was possible to adjust the hopper so that it was always upright whether the machine was travelling uphill or down. An improved form of this machine was in use in Norfolk.

5. An implement for dibbling potatoes. It consisted of a disc mounted on an axle and provided with projections at the rim. When it was dragged along the soil, each projection left an imprint into which a potato could be dropped. A single-row implement was intended for use by hand and a four-row dibbler was made to be drawn by one horse.

6. A barrow for broadcasting grass seed. A contemporary illustration shows a machine which is very similar to that used at the present day.

In the following ten years, 1830 to 1840, Crosskill introduced an improved turnip and manure drill, an implement for tearing the surface of grass land and broadcasting manure and grass seed, and a broadcast manure distributor.

The Suffolk drill, as made by Smyth, of Peasenhall, was an improved form of the old Norfolk drill, and was brought practically to its present form about 1860. The common practice at this time was to raise and lower the drill coulters by means of a chain and a roller set across the machine, and the coulters were forced into the soil by a similar method. The lever for raising the coulters and the maintenance of pressure on the coulters by means of springs was introduced about 1890.

In the eighteenth century there were two types of seed mechanism. The more primitive consisted of a small barrel or cylinder mounted on a horizontal shaft with a series of holes around its equator. As the barrel rotated seed which had been placed inside it fell out of the holes. The other type anticipated the modern force feed—a grooved wooden roller rotated in a cylinder at the bottom of the seed hopper, carrying a limited quantity of seed around in the grooves and dropping it through a hole in the lower side of the cylinder. Early in the nineteenth century small cups on short supports projecting from a horizontal roller carried the seed to the funnels, then later the cups were mounted on discs as in the modern Suffolk drill, and by 1880 all the modern feed devices were in use.

Modern corn drills are generally of one of two types. The Suffolk type is furnished with cup feed, rocking hopper, chain and roller lift and compression, hoe coulters and a special arrangement for steerage. This type of drill is drawn usually by more than two horses, driven and steered by one man while the drill adjustments are operated by another man. The other type of drill is more compact, having a lever lift, spring compression, force feed, and double or single disc coulters. It is driven and operated by one man and drawn by two horses. Root drills are usually provided with the cup feed and the older type of lift and compression, and are frequently made in the form of combined root and artificial manure drills. The variation in practice of sowing roots on the flat in some parts of the country and on the ridge in others, necessitates some slight variation in the design of the drills. Where roots are sown on the flat the drill is not usually provided with rollers, but for ridge work, a small concave roller leads the coulter and a smaller roller follows it. Seed barrows for broadcasting grass and clover seed are provided with a brush feed. They are made in a form similar to that in use in 1830 to be worked by hand, while larger implements are also made for horse work.

Manure Distributors.—The comparatively easy process of broadcasting artificial manures has become completely mechanized, although the very laborious work of distributing farmyard manure is still almost entirely done by hand. Farmyard manure distributors are being made by American firms and used in America for short straw manure, but although machines are on the English market which will deal with the long straw manure used in this country, the demand is very small.

It was probably recognized in the early days of artificial fertilizers that it was sound practice to place the manure in the row with the seed when root crops were being sown. This would give rise to the demand for combined seed and manure drills for sowing root crops, a demand which was being met quite early in the nineteenth century. Later, the principle of uniform distribution having been established, the broadcast drill for artificial manures followed and was in evidence by 1840.

The ideal manure distributor should be able to sow any artificial manure at any rate between $\frac{1}{2}$ cwt. per acre and 10 to 12 cwt. per acre, but there are many difficulties to be dealt with. Heavy vibration caused by travelling over rough arable land will make a drill sow more rapidly than it would do on a smooth pasture ; metal working parts which are in contact with the manure suffer corrosion and tend to clog with pasty manures such as superphosphate, whilst damp sticky manure may refuse to run through the drill, and the rate of sowing may vary as the drill goes up or down hill.

The simplest type of manure distributor is that in which the hopper is V-shaped. The upturned base of the V-shaped hopper is the lid through which the manure is introduced, and the hopper itself is divided transversely into two or three compartments to prevent the manure collecting at one side of the hopper when the machine is being used on the side of a hill. At the bottom of the hopper is a roller on which the manure rests. The roller is driven from the ground wheels, and, as the machine advances, rotates in the opposite direction to that of the wheels and carries a stream of manure out to fall to the ground. This kind of drill works well with dry powdery manures, but, in order that it shall be able to deal successfully with manures which tend to clog, a shaft is often mounted parallel to and above the roller inside the box, the shaft being fitted with short paddles which, by rotating, keep the manure agitated and so help it to run down freely to the roller. A different kind of distributor was imported from Central Europe before 1914. In this type there is the same transverse hopper, but in place of the roller there is a ledge and working along the ledge is an endless chain. The chain is provided with claws which project obliquely backwards so that manure resting on the ledge is pushed over the edge by the claws as they travel with the chain transversely across the machine. There are many variations of these two types, particularly of the former. A third type of distributor was patented in 1902. This consisted of a large hopper in the shape of an inverted truncated cone slung between the two ground wheels. From either side at the bottom of the hopper a shute led downwards ending just above a horizontal disc which rotated at high speed. The manure in the hopper was kept running to the shutes by a disc which rotated on the floor of the hopper. The manure passed down the shutes and on falling on the two discs was flung out on to the ground. With a capable man and a steady horse the distributor can be made to work very effectively in calm weather.

Another distributor is arranged so that the forward wall and the floor of the rectangular hopper rise slowly as the machine advances. A shaft is slung under the lid of the hopper and carries a number of radially mounted chisels. This shaft revolves, and as the manure is brought upwards by the rising floor of the hopper, the chisels scrape the manure backwards so that it falls over the rearward lip of the fixed side of the hopper. This has the advantage that there are no metal working parts in the manure and the mechanism is arranged so that the rate of distribution can be varied very widely.

RIDGING PLOUGHS AND HORSE-HOES

The Ridging Plough.—The ridging plough, used for ridging the land for root crops, has a similar Roman origin to the Kent plough, but it did not arrive in this country until the seventeenth century. Modern multiple horse-hoes are frequently made so that the hoe parts can be replaced by ridging bodies. Modern cultivators are also sometimes designed to take ridging bodies, but it is essential that any single or multiple ridging plough shall be capable of adequate steerage.

The horse-hoe was introduced by Jethro Tull early in the eighteenth century. The function of this implement is to aerate the upper surface of the soil, in order to prevent the evaporation of soil moisture and to uproot annual weeds. The horse-hoe is essentially an implement for the cultivation of the soil between the rows of the crop after it is sown, and so successful was this new departure that Tull formed the opinion that heavy crops could be grown by this method without the application of manure. The horse-hoe may be built for working one row at a time, or three or four rows at once for root-crops or six rows or more for corn crops. With multiple row horse-hoes it is always necessary to have some method of steerage, to deal with the same number of rows as the number laid by the drill, and when drilling the crop to drill the rows as straight as possible. The tines are usually of two types ; the A-shaped tine which takes the centre of the row and the L-shaped tines which are made left-handed and right-handed to take the sides of the rows, and in the case of root crops to permit cultivation of the soil as close up to the plants as is advisable. In horse-hoeing corn only the A-shaped tine is used.

Horse-hoes were being made in a variety of forms in 1830, and the two types of tine were already in existence, as well as multiple-row hoes and hoes with which were combined small drag harrows. The principle of the expanding hoe for easy and rapid adjustment to different widths was known, but its present-day form had not at that period been introduced.

HARVESTING MACHINERY

The sickle and the scythe are still in use for cutting grass and corn. In many districts the scythe is used for both purposes, but in some parts of the country the cutting of corn is usually done with the sickle rather than the scythe. The use of these implements, however, is now principally confined to the opening of a path for the binder around a field of corn, the harvesting of weather-beaten grain crops which cannot be dealt with mechanically, and the cutting of small areas of grass or green fodder.

Pliny, writing in the first century, describes a reaping machine, which would appear to be a more or less effective method of mitigating the toil of harvesting. This reaper was in use in Gaul, and it consisted of a cart, pushed by a single ox. On the forward edge of the cart fingers projected so that a kind of giant comb was run through the corn below the heads. The man walked beside the cart and swept the heads backwards so that they were cut and gathered in the cart ready to be pushed straight to the store or the threshing floor. This method of reaping by breaking off the heads is known as rippling, but in this country, where straw is needed for litter and for food, it has never been exploited.

An attempt was made in the eighteenth century to introduce the Gallic reaper into this country, and a machine for rippling corn, adapted from the design of the Gallic reaper, was brought out in 1787, but the idea was not adopted. An implement was devised in 1806 for rippling clover heads for use as seed and another was invented in 1826 for rippling beans, but neither became popular. The former was an adaptation of the Gallic reaper, while the latter was practically a plough with a rippling board instead of a mould-board, but even with such crops as these where the straw is useless the application of the principle attracted little or no attention.

In 1799 W. Walker, an author, in his " Familiar Philosophy in Twelve Lectures," described a reaper, but does not mention the inventor's name or any dates. The machine was pushed by a single horse, and it ran on two wheels connected by a live axle which carried a vertical bevel gear wheel. From the axle a framework was slung which lay close to the ground, and its forward end, which was toothed, was supported by a small wheel. On this framework a pulley was mounted horizontally below the axle, carrying above it a horizontal pinion which meshed with the gear wheel on the axle. This pulley drove by a cord a second horizontal pulley which was placed forward over the comb, and which carried a number of knives. Thus the cutting mechanism consisted of the comb and the horizontal rotating knives.

This machine apparently made no headway in practice, but it is interesting in its obvious derivation from the Gallic reaper, and in that

Bell's Reaping Machine (Cat. No. 99).

the principle involved, the horizontal rotating knife, seemed to obsess early inventors, even though it was incapable of extensive development. During the period 1799–1843 many inventors designed reapers on this principle. Some used the single rotating disc, others used a number of small discs, with adjacent discs rotating in opposite directions, others again built up a cylinder with a vertical axis and the horizontal blade as its base, the cylinder being intended to assist in carrying the cut corn out to the side to be laid in a swath. Smith of Deanstone constructed a machine on the cylinder principle which was still in use late in the nineteenth century, but for the most part none of these machines ever became a working proposition.

In 1812 John Common, of Alnwick, Northumberland, submitted a model of a reaper to the Royal Society of Arts. This machine had an angular knife reciprocating in a finger bar; the knife was driven through a connecting rod by an eccentric which was driven in its turn by a shaft which ran forward and which received its motion from a bevel gear wheel on a live axle carried between the two ground wheels. There is no doubt about the authenticity of this machine, though it received no premium from the Society because it had had insufficient trial. It is believed that Common suffered interference from the local peasantry, but whatever the cause, this machine, so exactly similar to the modern mower, which should have been regarded at the time as a great invention, was neglected even by its own inventor.

Ogle, also of Alnwick, with the help of Brown, the local smith, invented a machine in 1822, and Common claimed that his machine inspired Ogle. It is difficult to say whether Common's claim was justified or not. Brown is stated to have helped both men, and Ogle is said to have examined Common's machine. All three were neighbours, yet Ogle's machine had a smooth knife and a very different method of driving the knife. Ogle's machine also differed from Common's in having beaters and a hinged platform for sheafing mounted behind the knife-bar. Yet in an age when most men thought of rotating horizontal knives or discs and machines which for the most part are pushed, these two men designed two machines both of which were pulled, and both of which had a knife bar at the side with a reciprocating knife.

Neither of these machines, however, became a commercial proposition.

The first reaping machine that was ever used to a considerable extent was made in 1826 by the Rev. Patrick Bell of Carmyllie, Forfarshire. The original machine (No. 99) and a contemporary model (No. 100) are in the Museum Collections. The cutting mechanism shown on the actual machine is in the form of a reciprocating knife, but Bell's early attempts included a cutting mechanism similar in action to a pair of scissors (No. 100). Salmon of Woburn, in 1807, had also attempted a machine on the scissors principle, but the idea was not found practicable. Many trials of Bell's machine took place in 1828 and 1829, and in spite of the conservative attitude of farmers and active opposition of farm labourers, the machine gained ground in public favour. It came into use on farms quite widely, particularly in its own county of Forfarshire, but it was finally superseded by machines of American design during the middle of the nineteenth century.

In 1834 Cyrus H. McCormick patented in U.S.A. his first reaping machine, although the machine represented by the model (No. 101) is said to have been constructed and worked in Virginia in 1831. Further patents in 1845 and later were taken out (Nos. 102 and 103), and the McCormick reaper became a practical and a commercial proposition. At this distance of time, it is difficult to ascertain to what extent the prior designs of Common, Ogle and Bell assisted or stimulated the subsequent designs of McCormick.

Many reapers were shown at the Great Exhibition of 1851, but only four are worthy of notice. Smith of Deanstone exhibited a reaper which represented the highest possibility of the horizontal disc. The other three were those of Bell, McCormick and Hussey. The first very soon dropped out of the race, whereas the others in sundry trials up and down the land fought for supremacy for a number of years. McCormick's reaper and Hussey's were both American machines, similar to each other and to the modern reaping machine in general arrangement, but differing in detail. Bell's machine also dropped out, leaving McCormick and Hussey to produce finally a stabilized form of reaper which holds good to the present day.

There are numerous firms now making reapers and mowers, but there is little difference in general outline between the various makes. The mower for cutting grass is usually a two-wheeled machine which lays the cut grass in a swath directly behind the knife bar. The reaper has usually one driving wheel and a smaller supporting wheel at the far end of the knife bar. A platform behind the bar catches the corn as it falls, and serves as a sheafing table whence the sheaves are discharged sometimes backwards, sometimes to the side, either by hand or by means of some mechanism.

The evolution of the binder is an American story. The first step was taken by Jonathan Haines, who secured a U.S. patent in 1849 for a machine which was pushed by two horses and equipped with the Hussey cutting apparatus (the angular knife and finger bar). Behind the knife bar was an endless band, a canvas conveyor which carried the cut corn away and dropped it into a wagon which was drawn along by the side of the machine. This constitutes the introduction of the bed canvas and the lower elevator canvas.

S. S. Hurlbut, in 1851, secured a U.S. patent for a sheafing mechanism. He retained the bed canvas and lower elevator canvas, and allowed the latter to discharge the corn into a kind of trough. One may imagine a cylinder hollowed out so that its cross-section is Y-shaped, forming three troughs. This is hung with its axis parallel to the elevator rollers so that one trough is ready to receive the corn. As soon as it is carrying a certain weight the cylinder rotates throwing a sheaf to the ground and presenting another trough to be filled.

24

PLATE VII

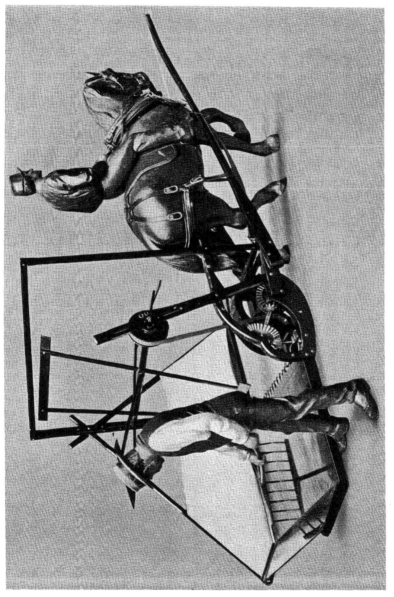

McCormick's Reaping Machine (Cat. No. 101).

PLATE VIII

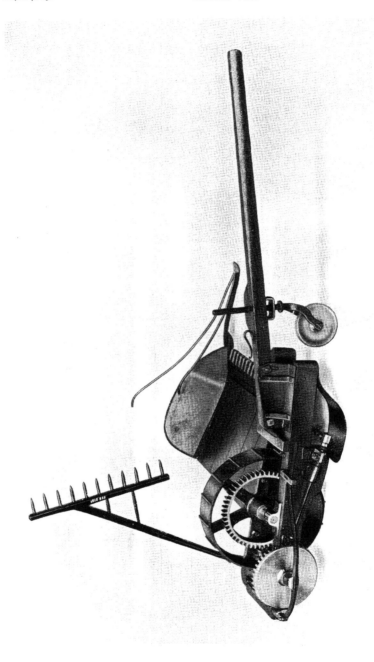

C. W. and W. W. Marsh, in 1858, obtained a U.S. patent for a machine that resembled the modern binder, except that in place of the tying mechanism there was a table at which a man stood (riding on the machine) and tied the corn by hand. In later patterns room was made for two and even three men. An automatic tyer using wire was invented by J. H. Gordon in 1874, and was sold as an attachment to the Marsh machine, and in the same year a similar machine made by the Walter A. Wood Mowing and Reaping Machine Company began to make headway.

In the same year, 1874, J. F. Appleby first worked on the twine knotter. He had patented a wire binder in 1869, and had been experimenting for several years, but the work begun in 1874 led up to the adoption of his twine knotter by the Deering Company in 1878. From this point onward the production of the modern binder was a matter of refinement of detail. The main problem was the reduction of weight and draught. British as well as American manufacturers have applied themselves to its solution, and by the use of high quality material for the parts of the machine, the provision of anti-friction devices, the improvement of the cutting mechanism, together with a high standard of workmanship, the modern binder, though quite a complicated, almost a delicate machine, is practically fool-proof, hard-wearing and of reasonable weight and draught.

Hay-making Machinery.—The process of hay-making entails the drying and curing of grass or clover after it is cut. With a light crop in dry weather the process may be complete in a few hours, but where the grass is juicy and the crop heavy so that the swaths left by the mowers are thick and bulky, it may be a matter of days before it is safe to carry the hay to the stack. Naturally the process will be prolonged if the weather is bad and through the farmer's anxiety to complete the operation, the hay may be put in the stack in a damp condition. The result of carrying hay too hastily will be excessive fermentation, often followed by spontaneous combustion and the loss of the crop by fire.

In hay-making, particularly in showery weather, it is essential to make the best speed possible when opportunity offers. For this reason and because the actual work, particularly the loading of hay, is laborious, a variety of machines have been devised and adopted for spreading and turning the grass and for collecting and stacking the hay.

The whole of the work was done by hand up to the end of the eighteenth century. The grass left by the scythe was turned or spread with rakes and forks, larger rakes were used to collect the cured hay into windrows, and the windrows were run up into cocks or the hay was pitched with forks into the cart, and by the same means it was pitched from the cart to the stack. In Scotland and the North of England, where rain was even more of a problem, the small cocks were collected together to make larger cocks or pikes and quite early in the history of hay-making in those districts a method was devised of roping the whole cock on to a low trolly, thus saving the labour of pitching hay in the field.

The tedder (No. 112), a machine for spreading the grass, made its appearance early in the nineteenth century. Salmon of Woburn patented such a machine in 1814, and described it as an open cylindrical frame mounted on the same axle as the two wheels. The frame revolved at a higher speed than the axle, and was driven by a pinion on the hub of one of the wheels which engaged with another pinion carried on a plate on the frame, this pinion driving the frame through an internal spur gear. The speed of rotation of the frame could be varied by replacing the second pinion by others of different sizes. On the frame, bars were fixed which carried the curved tines, each set of tines being spring mounted to neutralize possible damage through coming in contact with the ground. This type of tedder had become fairly common by 1830.

By 1890 the tedder had undergone considerable improvement. It had been provided with a double-action mechanism, so that the frame could be rotated backwards or forwards, and it was also made so that when the machine came to rest the tines continued to rotate for a short time, chiefly in order to allow them to clear themselves. A modified form of tedder (No. 113) had been introduced in which forks were mounted on levers and were made to imitate the tossing action of a fork used by a man. This was an attempt to devise something which treated the hay less violently than the forward action and more thoroughly than the backward action of the tedder. Horse forks and elevators had also been introduced in an attempt to minimize and accelerate the work of pitching the hay from the wagon to the rick. Pitching in the field from the ground to the wagon had received attention in the middle of the century. In 1852 a loader was patented, which was drawn behind the wagon. It was really a small and rather primitive elevator which picked up the hay from the ground, carried it up an inclined plane and dropped it in the wagon.

Toward the end of the nineteenth century the tedder seems to have come under a cloud. Whether the tines were rotated backwards or forwards its action was apt to be too violent, particularly so when the forward action was used. There was a tendency to bruise the grass and to thresh out grass and clover seeds, and where a crop of grass or clover was being cut for the purpose of saving the seed, this shortcoming of the tedder made its use prohibitive. It was also realized that unless the swath left by the mower was very thick and bulky, there was no need to spread the grass. On the contrary, curing went on quite satisfactorily in the swath if the grass was turned once or twice, rain did less harm to grass lying in the swath than when it was spread, and it was easier to make a windrow by rolling four swaths into one than by collecting spread hay with a horse-rake. Swath-turners which turned the swath over without breaking it began to be adopted about 1900, and once the principle of rolling the swath was adopted, it was not difficult to devise a machine which could act as a swath turner and with a suitable adjustment could take two swaths at a time and run them to one side to form a windrow.

26

Rakes were receiving attention early in the nineteenth century. The larger collecting rakes were sometimes mounted on wheels and were often fitted with shafts for use with a horse. By 1830 an implement for collecting hay had been brought from America (No. 117). This consisted of a wooden beam about ten feet long fitted with tines three or four feet long, a row of tines being carried on opposite sides of the beam, so forming a kind of double comb. The sweep was drawn by a single horse and controlled by a man through a pair of handles mounted mid-way along the beam. When the sweep was full the man raised the handles and the tines rotated on the beam, freeing the one set and bringing the second set into action. This implement is still used in some parts of the country, particularly in hilly districts.

The horse-rake had practically reached its present-day form in 1890, the tines were curved, and could be raised either by hand or automatically.

A small single-horse sweep was patented in 1903. It was somewhat similar to the earlier American sweep, but it had only one set of tines.

The large two-horse sweep made its appearance about the year 1902. This implement was practically a larger edition of the small sweep, with the horses attached close to the beam and at either end of it, so that the load collected on the tines between them. An interesting modern development of this implement has been achieved, in which horses are replaced by a tractor attached behind the sweep. This modification is very effective and assists in speeding up the process of carrying hay.

Potato Diggers.—Digging potatoes by hand is not only heavy work, but the slowness of the process makes it an expensive method of harvesting this crop. Where large areas were grown it is highly probable that quite early the plough was used to open up the ridges. A machine similar to the modern rotary digger was patented in 1855, although it was not the earliest attempt to invent a machine for this process, as a machine of the elevator type was patented in 1852. A modified form of plough was patented for potato digging in 1858.

At the present time potato ploughs are used to some extent for small areas, while the rotary digger is in fairly wide use. The rotary digger consists of a share which runs below the surface and lifts soil and potatoes, and behind it a spinner is set across the machine to rotate in the vertical plane. The mixture of soil and potatoes is sorted by the spinner, and the potatoes are thrown clear to lie on the surface of the soil. The older type of spinner was simply a number of tines protruding radially from a central hub, but it was found that the circular path of the ends of the tines was the cause of a certain lack of efficiency, through their being unable to reach potatoes at the sides of the rows, unless the share and spinner were set so deeply that the draught of the machine became prohibitive. This difficulty was overcome soon after the year 1900 by a device which caused the ends of the tines to follow a path which was an ellipse, the major axis of which was horizontal. This was a great improvement and some very effective machines have been made on this principle, but the improvement brought with it a disadvantage. The spinner which is made on this " Link " principle has a number of small bearings and moving parts within it which are liable to become clogged with earth or worn with grit. The obviation of this difficulty has been achieved by a recent innovation which is really the revival and improvement of an old nineteenth-century idea. The spinner is made without any moving parts, its plane of rotation is still at right angles to the line of travel, but it is tilted backward from the vertical so that, though the path followed by the ends of the tines is circular, the vertical projection of their path is an even flatter ellipse than that of the tines of the link type.

Potato diggers are also made which shovel up the mixture of potatoes and soil and carry it back over an elevator to a number of sieves. The soil is separated by falling through the web of the elevator, and the holes in the sieves, and the potatoes are dropped on the ground at the side of the machine.

Potato Sorters.—It is common practice in this country to separate the freshly lifted potatoes into three grades. The smallest potatoes, which are less than an inch in diameter, are classed as " Chats " and are used for feeding pigs. Those which are between an inch and an inch and three-quarters in diameter are saved for " Seed," to be sold as such or planted in the following Spring. All larger potatoes are classed as " Ware " and are intended for human consumption. The work of sorting is frequently carried out by hand, and the grading is done quite roughly by eye.

Machines for grading potatoes came into use late in the nineteenth century. In 1890 there was such a machine, which consisted of two sets of riddles, mounted so that they could be agitated by a man turning a handle. A second man fed the machine, while a third watched the potatoes as they fell on the upper riddle and removed stones, etc., and diseased potatoes. This machine was made in several sizes which were able to deal with 8 cwt. to 17 cwt. of potatoes per hour.

There are also machines of the elevator type which dig, sort, and bag the potatoes, but they suffer under the difficulties of high draught and inability to cope with poor soil conditions and stones.

At the present day mechanical sorters are used in those parts of the country where the crop is grown on a large scale. The general principle, however, of all types of potato sorters is very similar.

Flail, Piler, and Fan (Cat. No. 122).

Norag (Threshing Sledge) (Cat. Nos. 123–4).

THRESHING MACHINERY

The up-to-date thresher is not quite correctly named, as it not only threshes but also performs other quite important operations, such as the separation of the grain from the chaff, and the elimination of short bits of straw (cavings) and weed seeds from the main bulk of the grain. The true process of threshing is performed by rubbing the sheaves between the drum and the concave and is completed by passing the straw over the shakers. The next process consists of separating the cavings, the chaff and the weed seeds from the grain. This operation, known as winnowing, is performed by an arrangement of sieves and fans which comprise the parts of the thresher known as the first dresser and the second dresser. A further refinement is introduced at this stage when barley is being treated, when the awner and polisher are brought into action to remove the bristle which is carried on the husk of a grain of barley. A rotary screen performs the third operation. In this part of the thresher broken grains and small ill-nourished grains are separated from the main bulk to give an even, well-filled sample. This process of selection is carried out fairly rigorously—for example, where a good sample of malting barley is desired—and more rigorously still where it is intended to produce high quality seed corn.

Until the eighteenth century, these operations were performed entirely by hand. The sheaves were laid on the threshing floor and were beaten with a flail or threshel (No. 122), the straw being removed, leaving the grain mixed with chaff, etc. Where wheat straw was needed for thatching, the flail was often discarded, the man taking a double handful of corn from the sheaf, holding it by the butt ends and beating the heads on a stone or on a piece of iron shod timber. The awns were removed from barley by working the heap of grain with a piler (No. 122).

The flail is still used for small quantities of corn in some parts of the country, while abroad the time-honoured methods of treading out the corn with oxen, and of driving a sledge or roller over it are still in use.

Examples of the " Norag " used in Egypt (No. 123), and of the threshing board used in Cyprus (No. 125) are exhibited.

The separation of the grain from the chaff was done by means of an artificial draught created with the help of a fan, or by a natural breeze. In the latter case a sheet was spread on the top of a hillock, and the grain scooped up and allowed to fall five or six feet to the ground. The heavy grain fell back on to the sheet, while chaff and the lighter weed seeds were distributed far and wide on the wind. The fan was used generally under cover in the barn. It was a wooden roller with sacking nailed to it, mounted on a pair of rough bearings and turned by a handle (No. 122). As the fan turned, a second man with a scoop threw the grain in front of the fan, and as it fell the chaff was blown clear.

Prior to the introduction of sieves and screens the grading of the grain was carried out in very crude fashion. With a wide shovel, it was thrown across the barn floor into the draught of the fan, so that it fell in an elongated heap, the heaviest grains travelling farthest and the lightest forming a tail on the near side of the heap. In spite of the modern refinement of the process the name " tailings " is still retained for the shrivelled and broken grains.

As early as 1636 Sir John Christopher Van Berg patented a thresher. He was followed by Michael Menzies in 1734 and William Winlaw, an engine maker, in 1785, but none of these machines made any headway. The first practicable thresher was patented by Andrew Meikle, an engineer, in 1788. This machine seems to have been a modification of a machine used for scutching flax, and is no doubt the forerunner of the Scottish peg-drum thresher. Meikle in 1788, in Scotland, and Thomas Wigfull, a miller, in 1795, in England, both exploited the idea of combining with the thresher, apparatus for blowing away chaff and light corn, and separating with sieves the small corn and weed seeds from the main bulk of grain. A machine consisting of a plain drum with beater bars working against bars on a concave was patented by Palmer in 1798.

Threshing by machinery was apparently adopted almost with avidity in Scotland. Early in the nineteenth century such machines were reported as being common in that country, and were said to be spreading rapidly into England and Ireland. They were worked by hand, horses, water, wind, and to a small extent by steam. Stationary threshers were built into the barn, but portable threshers were also well known, particularly in the Midlands, and a portable steam thresher was in use in Northumberland. For the most part, winnowing was a separate operation performed by a hand machine which had been imported from Holland by Meikle and improved by him. By the middle of the century practically all the parts of the modern thresher had been invented and from then onward manufacturers have been occupied with improvement in detail.

A thresher can be made to last with ordinary care, particularly the simpler forms of the machine, for a good many years, so that at the beginning of the twentieth century in districts which were not very accessible, and where the adoption of all machinery was slow, there were still to be found small threshers, built in the barn. In most cases the winnowing had to be carried out by hand as a separate process, and the thresher itself was driven by horse-gear designed for four horses and sheltered by a " Round-house."

Prior to the War (1914–1918) the full-sized thresher with 54-inch smooth drum, double dressers, awner and polisher, and rotary screen, was the commonest form in use in this country, though the peg-drum was still common in Scotland. It was usually built to be driven by a

Threshing Board in Use, Vatili, Cyprus, 1929 (Cat. No. 125).

By courtesy of Dr. S. E. Chandler.
Winnowing at Vatili, Cyprus, 1929.

steam engine of about seven h.p., and was mounted on wheels so that the steam traction engine or horses could move it from rick to rick and farm to farm. Though the 54-inch thresher is still the standard form, a demand for threshers of large capacity, but of lighter build, has arisen since the introduction of the motor tractor, and this has been met by altering the design of the shakers, doing away with the elevator between the first and second dressers, and blowing the grain from the one part of the machine to the other.

Clover Hullers.—The process of separating clover seed from the pods, and the pods from the straw is essentially the same in principle as the threshing of grain crops, but the difficulty of breaking away the pods from the seed, and the smaller size of the seeds, give rise to certain differences in detail in the machines used. The general plan of a modern clover huller is somewhat similar to an ordinary thresher, except in the following details. An ordinary drum separates the bulk of the straw from the pods, whilst a mixture of unbroken pods, clover seeds, broken pods, bits of straw, weed seeds, etc., passes to a second drum in which the concave is replaced by a perforated sheet of steel. In this apparatus the seeds are separated from the pods and find their way through the perforations in the plate.

A certain amount of rubbish is held by the plate and is passed out to be shaken clear of seeds, but the mixture of small bits of straw, weed seeds, and clover seeds, which has passed through the plate, is taken with the shakings from the straw and passed over fine riddles and through air blasts. Thus by a process similar to that used in the first and second dressers of a thresher the clover seed is isolated, and graded.

CARTS AND WAGONS

The carrying of articles by human beings was the first form of transport, but as soon as man had obtained sufficient mastery over animals, the latter were brought into service. The load was either in a bag slung across the back of the animal or else carried in a pair of baskets or panniers hung one on each side.

These early methods are still employed in primitive countries. The sledge, by which heavier loads could be dragged along the ground, no doubt was evolved by a natural process of invention. Wheeled vehicles are known to have been in use from early Egyptian times.

For transporting agricultural produce a light two-wheeled cart was evolved, and is still used in the East. It consists of a body built up of planks or open rails fixed to a transverse wooden axletree, and provided with a pole or shafts for the attachment of the animals used for hauling it. While for larger loads the four-wheeled wagon was employed, it was found that four horses in separate carts could draw more than if attached to one wagon.

The types of carts and wagons used at the present time on the farm vary considerably from district to district. Such variations arise partly on account of the different purposes for which they are required and partly in connection with the hilliness or otherwise of the district concerned. Generally speaking, a cart is a two-wheeled and a wagon is a four-wheeled vehicle. The cart has much greater mobility than the wagon, and in the Midlands, for example, a big and heavy cart is used for all types of load, *e.g.* farmyard manure, sacks of grain, and lighter loads such as hay and straw. Such carts as these have, for some years, been constructed so that the body can be tipped backwards on the frame.

The four-wheeled wagon is used fairly generally in East Anglia, parts of the Midlands, Kent, Sussex, Hampshire, etc. The front wheels are smaller in diameter than the hind wheels in order to allow them to pass under the frame in turning. In this particular respect there is a good deal of variation, and the wheels must be made fairly small to give sufficient turning lock, but their reduction cannot be carried too far as a narrow diameter wheel on a loose surface gives a very high draught. Lades are usually fitted for use with bulky loads, and where a hay-loader is used in the field, and particularly where cart and loader are drawn by a tractor, high end and side lades are improvised if they are not fitted as a standard accessory. Two horses are needed for the wagon. An important part of the design of wagons and carts is the form of the wheel, both as to variations in the width of the rim depending upon the nature of the soil, and with convexity. The general tendency is, however, to standardize practice in design in various parts of the country.

BARN MACHINERY

Cake-breakers.—The use of oilcake for feeding cattle began in the latter part of the eighteenth century, and its introduction into Norfolk is one of the many benefits which Coke (1754–1842) is said to have conferred on that district. This food is supplied to the farmer in oblong slabs about an inch thick, and, particularly in the case of that which is made from cotton seed, the material is fairly hard. It was probably necessary from its first introduction to use some kind of machinery for breaking the slabs into fragments, and it is quite possible that machines similar to, or perhaps even the same machines as those for grinding bones, were used at first. Mention is made in 1830 of an oilcake bruiser, and this machine was a small edition of the contemporary bone-breaker ; it was worked by hand, and consisted of a pair of toothed rollers. This kind of machine apparently served the purpose throughout most of the century, for the first patent specification that bears any close resemblance to a description of the modern machine is dated 1888.

Chaff-cutters.—These machines are used for cutting straw and hay into short lengths of about an inch in order to effect economy in the use of such fodder. When straw is fed whole to cattle, they pick it over, and a considerable quantity is pulled down and trodden underfoot, so that where it has a high feeding value or when it is scarce, chaffed straw can be offered to the animal in smaller quantities and it is consumed more completely.

James Cooke, a clerk, in 1794, patented a knife, plate, weight and regulator for use with a machine for cutting hay and straw, and in his specification he refers to the " Common chaff-cutter frame." It appears that this frame consisted of a trough, rectangular in cross section, into which the straw was thrust, and along which it was moved by hand. As the straw reached the open end of the trough, it was either cut by a reaping hook or by a rotating wheel armed with blades. Cooke introduced a weight for compressing the straw at the delivery end of the box, and also fixed and rotating blades, the latter being attached to the spokes of a wheel. The fixed blade was attached to the end of the box and the rotating blades arranged so that a cutting action was obtained " similar to that of a pair of scissors." The weight and the scissor mechanism are characteristics of the modern chaff-cutter, so that it may be said that the machine as it stands to-day had its general principles laid down in 1794. In the following century various types of machines were introduced, but development seems to have taken place almost entirely along orthodox lines. An illustration published in 1831 shows the knives of a hand chaff-cutter to have a concave edge instead of the modern convex edge, but by 1890 the convex knife had arrived and machines were adapted for use

c

with horse-gear. The large chaff-cutter which is fed automatically straight from the thresher had also made its appearance. Alteration of the length of chaff cut is effected by altering the speed of the feed rollers relative to the speed of the knives. This improvement was introduced in 1865, but in 1873 a further improvement was introduced. The modern clutch was invented whereby it became possible to stop the feed rollers or to reverse their motion while the machine was running, in order to prevent accidents through the operator's arms being drawn into the knives. The illustrations to the 1865 and 1873 patent specifications show chaff-cutters which are similar to the modern machine, indicating that the design of these machines was stabilized soon after the middle of the nineteenth century.

A modern development of the chaff-cutter is the silage-cutter and blower. It is found that green stuff packs away more easily, and that the process of ensilage * is better under control when the crop is cut into short lengths. For this purpose the chaff cutter has been pressed into service, and slight modifications have been introduced. The speed of the feed rollers is rather higher as it is not necessary to cut green stuff as short as straw-chaff, and the upper compressor roller is made to react against springs instead of against the usual weighted lever. It was found convenient to stack silage in tall, narrow towers, and in order to be able to feed the tower from the ground a blowing mechanism was added to the cutter. The cutting wheel is completely enclosed and paddles are fitted to the rim of the wheel thus converting the cutting wheel into a fan enclosed in a fan case. To the case is attached a pipe which is carried without any bends to the top of the tower, where it curves over, thus the cut material is immediately blown up the pipe and so into the tower.

Root-cutters.—Turnips, swedes and mangolds are the roots which are most generally fed to stock on the farm. No doubt, at first the roots were fed whole, but by the beginning of the nineteenth century the practice of slicing roots was very common. Not only can fuller use be made of a sliced root by the animal, but certain special circumstances occur which make slicing a necessity. Young cattle and sheep go through a stage when they lose their first teeth and grow a permanent set, and during this period they are unable to deal effectively with a hard, whole root. It was also discovered that if sliced or pulped roots were mixed with chaffed straw and allowed to stand, the fermented mixture was more palatable to the animal.

The earliest root-slicers were similar in appearance to a barley-piler, except that the knives were further apart. The root was placed on the ground and the slicer was dropped on it sharply, so cutting the root to pieces. Another form of slicer consisted of two hoe-like blades set at right angles to each other and fixed to the back of a fork. The root was pulled clear of the heap with the fork, and was then cut by the inverted implement. The implement was, no doubt, very effective.

* Ensilage is a method of storing green fodder for preservation, by storing in towers or pits.

34

In the early nineteenth century a slicer appeared which consisted of a lever hinged to a trestle table. Knives were fitted in a hole in the table, and the root was pressed down on to the knives by the lever. What may be called rotary root-cutters made their appearance between 1830 and 1855.

Root-cutters which can be worked by turning a handle, and can be adapted to be driven from a counter-shaft or an engine, can be divided into three classes : slicers, cutters, and pulpers. The slicer cuts the root into fairly large portions, the cutter reduces the root to fingers which are square in cross section and sometimes as much as an inch thick, while the pulper reduces the root to very small pieces.

Sheep-Dippers.—Sheep-dipping should not be confused with sheep-washing. The latter is practised in order to clean the wool of the sheep prior to shearing, and is a custom which is confined to certain localities. When sheep are dipped they are immersed in a disinfectant in order to destroy parasites, and as a precaution against the contagious disease " Sheep-scab."

Apparently scab was not troublesome as a disease of sheep until the end of the thirteenth century, and from then until the nineteenth century, no concerted measures of prevention were adopted ; but the more careful farmers used tar for direct application to the affected parts of the sheep. The use of a disinfectant lotion seems to have begun early in the nineteenth century. At any rate the use of baths for dipping is indicated by the submission of a model bath to the Highland Society in 1857.

The apparatus used at the present day falls into two classes, the portable wooden bath and the swim bath. The former represents the older method, and the use of the portable bath is rapidly becoming confined to remote districts and small flocks. The apparatus consists of a wooden bath big enough to contain a sheep, and sufficient liquid for practically complete immersion. Alongside the bath is a sloping table on which the sheep is placed on its back while the lotion is rubbed into every part of the fleece. There is also a wooden platform on which six sheep can stand while the surplus lotion drips from them and runs back into the bath. This method entails considerable lifting and manipulation of the sheep, and is also wasteful of the quite expensive lotion, so that where large numbers of sheep are to be handled a less laborious and expensive method was desirable. To meet these requirements the swim bath was introduced, late in the nineteenth century ; it was known, but its use was not widespread, in 1890. In this case the bath is larger, and is sunk into the ground so that its rim is level with the concrete flooring of the pens at either end of it. It is also made long and narrow, so that one sheep can be dropped in at one end while another climbs up the slope at the other end, but there is not sufficient width for a sheep to turn round. The animal has to swim a short distance, and in doing so its actions ensure that the disinfectant is worked into every part of the fleece. A collecting pen is laid out at the near end of the bath, the pen being sufficiently small for ease in catching the sheep, and draining pens, usually two, are provided at the far end of the bath. The draining pens have a concrete sloping floor, and are large enough to hold about twenty sheep each.

One man, with one or two helpers, catches the sheep and tips them into the bath at the near end ; a second man manipulates a gate which keeps the sheep (treading water) in the bath about one minute, and he also with the aid of a crutch pushes the sheep completely under once or twice. The sheep is then allowed to walk out of the bath into one of the draining pens. First one draining pen is filled with its twenty sheep, and they are kept there until the other pen is full, so giving time for surplus lotion to drip from the fleeces and to trickle back into the bath.

Shearing Machines.—A machine for shearing sheep, which was hailed as a promising invention, was in use in Australia about 1890. It consisted of a cutting wheel geared to the shaft of a small steam turbine, worked by a current of steam conveyed from the boiler by an india-rubber tube. A comb was fitted to the cutter to avoid injury to the sheep, and the working parts and hand-piece were made of brass. The clipping was said to have been done rapidly and with safety to the sheep.

The older method of shearing was by means of the hand-shears, and the introduction of the machine allowed the shearer's hand to travel more quickly and a wider swath of wool to be cut at each stroke.

The modern hand piece (No. 185) consists of a steel comb on which a steel knife reciprocates. The motion of the knife is obtained from a pulley on an engine or a lay shaft by means of a flexible shaft running from the hand-piece to a pulley which is connected with the prime mover. A single hand-piece may be driven from a wheel turned by hand—a method often used in clipping horses—or from an engine-driven pulley. Where several hand-pieces are to be used a number of bevelled friction pulleys are mounted on a single shaft, one pulley for each hand-piece, and the shaft is driven by a petrol or paraffin engine or an electric motor.

MILLING MACHINERY

Crushing and Grinding Mills.—Grain, before it is fed to stock, may be treated mechanically in one of three ways. It may be rolled, kibbled or ground. Rolling is the treatment generally applied to oats, and it consists of applying just sufficient pressure with a smooth roller, to burst the husk of the oat and expose the kernel. By this means, whether the horse that eats the oat chews thoroughly or bolts its food carelessly, there is more chance of the food being well digested and less possibility of wastage.

The same principle applies to beans and to barley when these foods are used, but a rather different apparatus is required to crack the bean or the barley grain. Oats are allowed to run between two smooth rollers, but to deal with beans or barley it is necessary to use a machine built on the principle of the mill. In this apparatus a rotating plate revolves in close promimity to a fixed plate, and the two adjacent surfaces consist of roughened stone or metal. The difference between kibbling and grinding is merely one of degree. A pig bolts its food, and so barley meal which is to be mixed with wash to make pig food, needs to be ground finely. For such a purpose the plates of the mill are set very close to each other and the barley is fed into the mill so that it passes through slowly.

On the other hand, cows and sheep chew their food much more thoroughly, so that it is not necessary to grind their food so finely. There is also the danger that a finely ground meal which is fed in a dry state to the animal may be wasted by being blown away. In these circumstances the mill plates are set fairly widely apart, and the grain is run through rapidly thus obtaining just sufficient treatment to break the individual grains, and to obtain quite a coarse product.

The stone mill for grinding wheat is almost as old an agricultural implement as the plough, and it is difficult to say how or when the practice of grinding other grain for stock feeding may have crept in. The use of cast metal rollers for " crushing, flattening, bruising or grinding of malt, oats and beans," etc., was patented by Isaac Wilkinson, of Lancaster, in 1753. The introduction of cast iron in the manufacture of these machines would facilitate the construction of compact machines needing less driving power, but apparently little else than the roller type was in use in this country in 1830. It is described as being suitable for kibbling grain and beans and also for grinding malt.

The most important inventions, which led up to the types of machines which are in use to-day, were brought out in the following order. The use of cast iron discs to replace the clumsier stones of the grinding mill commenced in about 1855. The use of conical surfaces instead of flat surfaces to increase the grinding area without increasing too greatly the diameter of the stones was being exploited in 1860. The mounting of the grinding units on a horizontal instead of a vertical axis is mentioned in a patent specification in 1866. A patent was obtained by Bamford in 1886 for reversible grinding plates in disc mills. The modern feed device found on a Bentall mill was patented in 1898.

A considerable amount of work was done during the middle of the nineteenth century in attempting to perfect the " Roller and breast mill." In this machine a fluted or serrated roller on a horizontal axis was made to work against a curved breast which was also fluted, and which was adjustable to and from the roller. The inventions mentioned above led up to the modern mill in which the grinding discs are mounted on a horizontal axis, are made of cast iron, are sometimes flat and sometimes conical, and are provided with replaceable grinding surfaces. This latter type of mill has been developed into a compact, highly efficient machine, and for agricultural purposes has completely superseded the roller and breast pattern.

Flour Milling.—The earliest method of reducing grain to flour was by means of a pestle and mortar, represented by a pair of stones, concave and convex, or by a hollowed log and a club. The hand quern succeeded these and acted like our present mill-stones by a circular grinding action ; increased in size and driven by power, the nether millstone became our " bedstone," and the upper one the " runner " ; in some modern mills, however, the stationary stone is the upper one. The meal from the stones is usually dressed or sifted through screens so as to remove the bran which in passing between the stones is only reduced to flakes of comparatively large size.

Since about 1890 the use of millstones in the production of the whitest flour has been almost entirely superseded by roller-milling, and the gradual reduction process, in which the grain is reduced to flour between rollers by a series of stages. The grain is passed successively between pairs of chilled, cast iron rollers, of about 10 in. diam. by 30 in. long. Each pair is set closer together than the preceding pair, and after each passage or cracking the product is screened or " bolted " so as to separate the floor reduced. In this way the flour from the successive layers of a grain of wheat are obtained considerably classified, and the market value of the highest qualities is such as to render the separation profitable.

In all mills the flour dust in the air will burn explosively under certain conditions, so that the collection and removal of this dust becomes an important matter. Settling rooms have generally been employed, but a more recent dust separator is seen in No. 186.

Essex Mill (Cat. No. 192).

DAIRY MACHINERY

Milking Machines.—Attempts have been made over a number of years to invent machinery which will replace the hand milker. There are at the present time a number of different makes of milking machines, which are stated to milk efficiently, without damage to the cow. With all makes of machines, however, the operator must draw off the last few ounces of milk by hand. Recent tests have shown that if the machine is properly cleaned, the keeping quality of machine-milked milk need not be inferior to that drawn by the best hand-milkers.

In the nineteenth century several attempts were made to expedite the process of milking by the insertion of tubes into the teat. Progress along this line was not possible owing chiefly to the difficulty of keeping the tubes clean and to the danger of setting up and spreading disease of the udder by the use of dirty tubes. About the middle of the century it was realized that in some way the pressing and sucking action of a calf's tongue and palate must be simulated, and attempts were made to design a machine for this purpose.

Various inventions at this period led to a type of machine which is in operation successfully at the present day. This machine comprises a suction pump, driven by an internal combustion engine, and connected by a flexible air-tight tube to an air-tight chamber in the lid of the milk bucket. The chamber in the lid possesses a valve which closes and remains closed when the pressure in the chamber falls below atmospheric pressure, and the chamber is also connected to the four metal teat-cups which are fitted with rubber rims and linings. When the piston of the pump moves outwards the pressure in the lid chamber and teat cups falls, and milk is drawn from the udder and collects in the lid chamber. On the return stroke of the piston the pressure in the system rises, the suction on the teats ceases, and the opening of the valve in the lid chamber allows the milk which accumulated during the previous suction stroke to fall into the bucket.

The modern milking machine (No. 203) usually consists of pipes laid along the cow-stalls above the heads of the cows with a bail tap between each pair of cows, the pipes being connected to a vacuum reservoir which is exhausted by a mechanically driven pump. Pressure gauges and safety valves are also common to most makes, and there is an airtight pail, which attaches to the bail-tap by one flexible tube and to the four teat-cups by another. The pulsator, the object of which is periodically to break the vacuum and so imitate the alternate suck and release, characteristic of the calf's action, is the most variable component both in its form and its position in the system. In most cases it is found on the lid of the pail which receives the milk, although in one machine there is a small pulsator at the base of each teat-cup, and in another the pulsator is carried in the small metal tube which forms the junction of the four tubes coming from the teat-cups. The last machine is a modern machine which has a double system of vacuum tubes, and attempts to produce in this way suction and mechanical pressure. Other designs rely on the small amount of mechanical pressure which will accrue when a rubber sleeve inside a metal teat-cup is fitted over the teat.

The most important problems that remain are probably the simplification and perfection of the process of adequate cleaning, and the accurate fitting of teat-cups.

Cream Separators.—Milk can be split up without great difficulty into the heavier liquid—the water with its soluble matter and the casein—and the lighter material, the butter fat or cream.

The removal of the butter fat globules is required if cream or butter is needed, and to do so the fact is exploited that they are lighter than the liquid in which they float. Until the later years of the nineteenth century there was no other known way of separating cream than the gravitational method. The raw milk was allowed to stand for twenty-four hours, and the cream formed a layer on the surface. It was then skimmed off by some means such as a perforated, saucer-shaped metal scoop.

Centrifugal separators had been in use in various branches of commerce for many years, but not until after 1877 were there any serious attempts to apply centrifugal force to cream separation. A patent was registered in that year covering an apparatus consisting of a wheel mounted on a vertical spindle, to the periphery of which buckets of milk were attached. Rotation of the wheel entailed the spinning round of the buckets. The milk which they contained was subjected no longer to gravity alone, but also to another force tending to pull the non-fatty constituents away from the central spindle and allowing the layer of cream to form much more rapidly.

The principle involved was applied in a more practical manner in 1878 in a patent registered in this country by Alexander for Lefeldt and Leutsch. The apparatus was similar on broad lines to the modern separator, consisting as it did of a bowl mounted at the top of a vertical spindle. Milk was poured in from above at the centre of the bowl while it was revolving, and skim milk was drawn off into a tray below the bottom of the bowl, while cream which collected at the surface was allowed to run over the lip of the bowl to be caught in a second tray.

Dr. Gustave De Laval, also in 1878, invented a machine (No. 204) which was more efficient than that of Lefeldt and Leutsch. It was similar in principle, but differed in many important details, the chief of which was that it was a continuous discharge separator, the machine not having to be stopped to draw off the skim milk. It is stated to have been the first machine of this kind.

It is very similar in outline to the modern machine, but the similarity is even greater in the separator patented by him in 1887. There were, however, certain elements in the construction of the bowl, and an arrangement for the regulation of the raw milk feed needed, before all the parts of the modern separator could be said to be present. There has also been since that date on the part of all makers, great improvement in the details of design and in materials used for the construction of separators.

The hollow bowl of the early separator did not ensure that all the particles in the liquid were being spun at the same rotary speed, and there was also an excessive amount of frothing with the cream and separated milk discharged. To overcome this, Bechtolsheim in 1889 introduced a bowl the interior of which was divided up by a number of thin conical partitions placed a little apart throughout the height of the bowl. Although Bechtolsheim conceived the idea of inserting conical discs in the bowl, this improvement was not exploited until De Laval in 1890 introduced a new type of separator with conical discs. Lister and Pedersen in 1892 patented a separator in which the raw milk was introduced from a tap into a tray which corresponded to a dilatation of the upper end of de Laval's vertical raw milk tube. In the tray was a float which when the milk reached a certain depth, closed the tap, and so regulated the rate at which the milk was allowed to run into the bowl. The modern conical bowl is first seen as an invention patented by Strom in 1894. It would therefore appear that the form of the separator was definitely established in the early nineties, and there is evidence that by that time it was already in use to a limited extent on farms in this country. Since that date its efficiency has been increased, and its popularity has grown until there is hardly a farm where milk is produced which does not possess a modern type of separator.

Butter Churns.—The properties of a good churn are, that it should be of such a form internally that the whole liquid mass of cream can be equally and thoroughly agitated ; that its solid contents after churning (the butter) can be easily removed ; and that it can be easily cleaned, having no crevices inside where dirt can lodge or escape observation. The lid must fit firmly, forming an airtight joint, there should be a means of ventilating the churn, and a pane of glass through which the state of the contents can be observed, and it should be so designed that the labour of working it is not too great. Primitive methods of churning consisted of either agitation of the vessel containing the cream or the agitation of the liquid in the vessel by means of a plunger. Since the middle of the nineteenth century the number of different varieties of churns has increased, so that Stephens, in his " Book of the Farm " (1891), says that " The dairy farmer may now have almost as much scope and freedom in selecting a churn as in choosing a wife." Out of the many types which have been submitted to the test of prolonged use the one which has held the field and is still the most popular is the end-over-end barrel churn (No. 216). This churn is built in much the same way as an ordinary cask except that one end is removable and forms the lid. It is mounted eccentrically about its middle part so that it turns end over end, and is usually turned by hand. The wide opening, when the lid is removed, makes it easy to take out the butter, and the absence of beaters or internal flanges facilitates cleaning. The glass pane and blow-off valve are usually carried in the lid, which is held securely by four or six clamps.

Cheese Presses.—The making of a hard cheese such as a Cheddar is a fairly complicated process, but one of the important operations involved in the later stages is the extraction of the last quantities of whey from the curd. This is done by placing the salted curd in a chessart or cylindrical mould, open at both ends and applying pressure through the open ends to the extent of two to three tons weight. The early presses were made having a rigid platform on which the cheese was placed and the weight was applied to the cheese by placing weights on a super-imposed platform which pressed down on to a cylindrical plunger fitting into the mould. Such a method entailing a clumsy apparatus and the lifting of heavy weights by the operator encouraged the adoption of presses which were made on the screw principle. In the early stages screw presses failed because the pressure was not maintained when the mass of the cheese decreased in volume, and attempts followed to combine a falling weight with the screw principle. A primitive solution of the problem is seen in Brown's Patent press (No. 222), while the principle of the modern press is illustrated by a model (No. 224). In the modern press the plunger can be screwed down on to the cheese quite easily, and pressure is applied and maintained by the system of weights and levers. This arrangement was embodied in a patent covering an invention by George Travis, of Derby, in 1857, and the general form of the cheese-press has altered little since that date, though there has been considerable improvement in detail.

The assistance of H. G. Richardson, Esq., and his colleagues in the Ministry of Agriculture, is gratefully acknowledged.

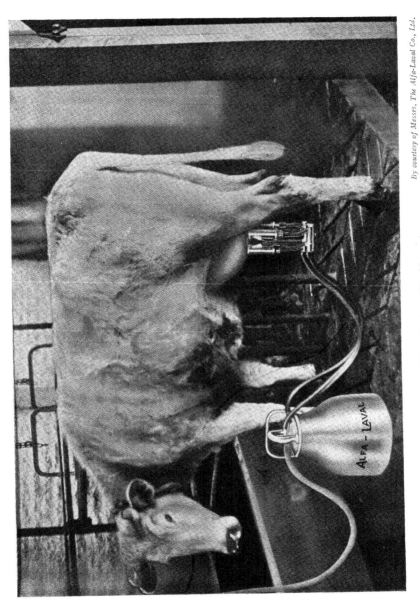

By courtesy of Messrs. The Alfa-Laval Co., Ltd.

Milking Machine (Cat. No. 203).

TILLAGE IMPLEMENTS

1. STAG-HORN PICK. Presented by Commander E. C. Shankland.

This is a fragment of either a pick or hoe fashioned out of the horn of deer. Such implements were used in this country by primitive man.

The hole near the root of the horn is probably intended to receive a wooden handle, and the implement was used perhaps single-handed like a hoe for stirring the soil. It could not have been very effective. The fragment was found in the Thames near Wandsworth in 1928.   Inv. 1929–970.

2. CASCHROM, *i.e.* CROOKED FOOT DIGGER. Presented by John McInnes, Esq.

Implements of this type were in use in the Hebrides and Isle of Skye even as late as 1860, by those who could not afford horses or oxen for ploughing.

When in use the handle rests on the ploughman's shoulder and he pushes forward the digger with a double jerk of his foot. The clod thus separated is thrown from left to right, the labourer taking a step backwards after turning each successive clod. It is stated that the rate of progress of an expert is about one-tenth of a Scottish acre per day. The implement acts like a narrow spade digging a groove the depth of which is determined by raising or lowering the handle.   Inv. 1913–497.

3. EARLY FOOT DIGGER. (Scale 1 : 16.) Lent by Major Steinmetz.

This represents the crooked foot digger, or Caschrom, used as late as the nineteenth century in some parts of the Hebrides.   Inv. 1926–823.

4. BREAST PLOUGH. Presented by Mrs. W. B. Burt.

In the eighteenth century and previously the breast plough was practically the only implement available for skimming turf or rubbish from the surface of the soil. The plough is worked by one man who pushes it forward in short stretches. The specimen shown was obtained from Wellingore, in Lincolnshire, and a similar implement was seen in use in 1926 at Quedgely, in Gloucester.   Inv. 1926–629.

5. ETCHING OF A BREAST PLOUGH. Presented by H. G. Richardson, Esq.

This picture illustrates the breast plough in use.   Inv. 1926–995.

BRITISH PLOUGHS

6. RANDELL'S NORFOLK PLOUGH. Presented by the Castle Museum, Norwich.

This plough was made in Norfolk, possibly in the latter part of the eighteenth century, but it embodies features which date back to the middle of the sixteenth century.

The high angle of the beam is a common feature of Norfolk ploughs, and is supposed to throw pressure on the heel of the plough, preventing any tendency for it to ride on the share-point. The single handle is also a distinctive characteristic, and may be a legacy from the Dutch. There are fittings on the beam for a skim coulter and a knife coulter, the former being at that time (middle to late eighteenth century) a novelty. Other features which were new at that time are the iron frame which is a replica of the old rectangular timber frame ; the slade, which is bolted on horizontally below the frame ; and the cast iron share, which fits on to the fore end of the slade. The mould-board is what would now be called a general purpose type, being neither long enough for good autumn ploughing nor sufficiently bluff for good digging work.   Inv. 1929–554.

7. HOOK'S NORFOLK PLOUGH. Presented by the Castle Museum, Norwich.

This plough may have been made at much about the same time as Randell's plough, that is, either in the late eighteenth century or the early nineteenth century.

It is similar in many respects to Randell's plough with the following fresh features.

There is a curious arrangement for the adjustment of the pitch of the knife coulter ; a well-designed skim coulter is included ; the mould-board stay, which is attached to the hinder part of the frame and holds the mould-board out across the furrow, is adjustable ; and a second handle has been added. As in Randell's plough the width of the furrow-slice can be adjusted at the draw-bar (or wilds) on the fore-carriage and at the tow-chain shackle, which is situated on the beam between the two coulters. Depth of ploughing is adjusted by raising or lowering the pillow—the cross-bar of the fore-carriage on which the beam rests. Inv. 1929–553.

8. EARLY ENGLISH PLOUGH ; ROTHERHAM TYPE. Presented by John McInnes, Esq.

The plough shown is a very good specimen, probably very similar in appearance to the original Rotherham plough. As originally manufactured at Rotherham in Yorkshire, the plough had two iron shares, the second fitting above the ordinary share so that it shielded the breast, the mould-board was curved and was made of wood covered by two iron plates. The example shown also has a wrought iron share, the second share and the mould-board plates being absent. The coulter blade and the slade bottom are of steel. Inv. 1913–496.

9. ROTHERHAM PLOUGH. (Scale 1 : 16.) Lent by Major Steinmetz.

The plough represented marks a great advance in soil cultivation, for it was one of the first ploughs with a curved mould-board to be used in England. It was patented by Disney Stanyforth and Joseph Foljambe in 1730. The knife or coulter fixed to the shaft cuts the soil for the passage of the share. The curved mould-board, which was later made of sheet iron, turns the furrow over. Inv. 1926–831.

10. TWO SHARES. Presented by Messrs. Ransomes, Sims and Jefferies, Ltd.

The two shares are cast from early nineteenth century patterns and are typical of the kind of share used between 1800 and 1820. Both shares show a closer relationship to Arbuthnot's share described in Young's " Eastern Tours," Volume 2 (1771), than they do to the modern lea share. They differ from the modern share in having greater height at the hind end of the land side, and also in the way the land-side edge leads up to the breast of the plough.

The upper edge of the land-side of the share of a modern plough is a straight line, whereas that of the two shares shown is the beginning of an ellipse very similar in form to Arbuthnot's share. Inv. 1927–1998.

11. SOMERSETSHIRE PLOUGH.

This plough is said to have been in use in Somerset within the last forty years. Its design is characteristic of the mid-eighteenth century, but it was probably built at a much later date.

The iron share with its fin and its continuation backward to the heel, is a feature which is strongly reminiscent of the share described by Jethro Tull and Hale early in the eighteenth century in connection with the improved Hertfordshire plough. This share differs from the Hertfordshire share in that the breast is also welded to its upper surface.

The mould-board is an example of the early winding board, which came from the Netherlands to East Anglia in the seventeenth century. A piece of timber was shaped roughly to the natural turn of the slice and was then scoured to its correct shape by constant use. In this case the old ground-wrist is no longer a separate part, but is embodied in the mould-board itself, and as is the case in a modern mould-board the ground-wrist is represented by a small piece of iron fastened to the underside of the board to protect it from excessive wear. Inv. 1928–194.

12. DERBY PLOUGH. (Replica copied from an original plough in the Museum, Torquay.)

This represents a small general purpose wheel plough, built in the early half or the middle of the nineteenth century, and the type was probably in use until the end of that century.

The example shown is probably typical of the light effective implements which arose as improved forms of the Rotherham plough and there is no difficulty in realizing why their value was so great in comparison with the older clumsier varieties.

Adjustment for varying the width of the furrow slice was carried out by moving the draught rod from one hole to another laterally in the shackle.

Variation of the depth of the furrow could be effected in three ways, at the lever neck, at the head wheel and at the point of attachment of the shackle to the beam. Inv. 1927–996.

44

13. HOWARD'S PATENT H.H. WHEEL PLOUGH. (Scale 1 : 4.) Lent by The Royal Scottish Museum.

This type of plough was made by Messrs. Howard, of Bedford, and was sent to Scotland some time before the year 1865.

The share is broad and flat and is held well forward, while the breast runs back from it without much rise and shades into the curve of the mould-board which twists gradually with the natural turn of the furrow slice, bearing against the slice as uniformly as possible as it turns along its length. Thus the body of the plough is designed to do good autumn work, to cut a rectangular slice and twist it through a right angle and a half and to lay it unbroken against the previous slice. Inv. 1928–400.

14. HOWARD'S CHAMPION PLOUGH. (Scale 1 : 18.) Lent by H.M. The King.

This model was presented to His Majesty King Edward VII. when Prince of Wales, on the occasion of his visit to the Dublin Agricultural Show, in 1868.

It depicts in detail a long-plate plough such as is now used at ploughing matches, and the fact that it is sixty years old, not only shows how little improvement in detail has been necessary in the design of ploughs since the middle of the nineteenth century, but a comparison of this plough with Howard's own T.A. plough (1820) shows how rapid was the development in plough designing in the early half of that century.

The model was made by Messrs. J. and F. Howard, Ltd., of Bedford, to represent their " Champion " plough. The frame is similar to the modern frame, and is complete with a lever neck, for adjustment of vertical pitch. On the landside (left) of the frame is fitted the slade, on which the plough slides and the side-cap which presses against the wall of the furrow. The edge of the share is slightly concave so that it would bear on a level surface only on the share point and the point at the furrow side (right) end of the edge. The breast (the line running from the point of the share up to the frame) is very long and the surfaces of the share and of the fore part of the mould-board are much more nearly horizontal than they were in ploughs made earlier in the century. The mould-board itself is also long and twists gradually. The slice is cut vertically by the coulter and horizontally by the share, and is pivoted uniformly on its furrow side edges, as the breast slides underneath it. The mould-board completes the process, finally laying it at an angle of 45 degrees to the horizontal and pressing it against the preceding slice.

On the beam, ahead of the coulter, is fastened the skim coulter. This is practically a miniature plough body which turns a tiny furrow slice, the landside of its furrow coinciding with the line of the coulter. In this way turf or manure lying on the surface of the soil is completely buried instead of protruding between the upturned slices.

In this plough the draught is taken by a rod which attaches to the beam near the frame and which runs forward under the beam. The hake at the fore-end of the beam allows the draught rod to be raised or lowered and the line of draught to be adjusted to right or left of the general line of the plough. Inv. 1928–245.

15. MODERN WHEEL PLOUGH. (Scale 1 : 8.) Lent by Major Steinmetz.

This is a model of a twentieth century wheel plough of the type made by Messrs. J. and F. Howard, Ltd., of Bedford, for autumn work. Inv. 1926–832.

16. LONG-PLATE PLOUGH. Lent by Messrs. J. and F. Howard, Ltd.

This is an example of the common English plough of the present day.

This type of plough receives its name on account of the length of the mould-board, a feature which is characteristic of English ploughs intended for the English practice of autumn ploughing. Inv. 1928–994.

17. DIGGING PLOUGH. Lent by Messrs. J. and F. Howard, Ltd.

This is a modern plough used in the spring for " Digging work."

In this plough the share and breast have a short bluff shape so that as the slice slides on to the mould-board, it curves and tends to crumble at once. This effect is increased by the shape of the mould-board itself, which is shorter, and twists more sharply than that of the long-plate. The tail of the mould-board tends to throw the slice down and to allow it to pulverize itself still further as it falls. The bar attached to the tail of the mould-board increases this effect and even greater pulverization is sometimes achieved by replacing the bar with two or three knives set at an angle so that the slice is thrown against them before it falls. Inv. 1928–995.

45

18. TWO-FURROW PLOUGH. (Scale 1 : 4.) Lent by Messrs. J. C. and T. Yates.

This is a model of a double-furrow plough, by which implement one man is able to perform twice the amount of work that he could with a single plough.

The model shows a plough with a timber beam strengthened with wrought iron, and provided with a notched bar in front so that the draught may be exerted exactly in line with the combined resistances of the two shares, and the slice be maintained of constant width. By carrying the body of the after plough by slotted or adjustable brackets the width of the furrows can be varied from 7½ in. to 11 in. apart. To lift the plough, a hand lever is employed which moves two arms, one carrying a wheel and the other a shoe, the wheel and shoe being simultaneously depressed by pulling the lever. By these means the plough is lifted out of the ground, and its weight carried on the two wheels and the shoe. The front wheel is adjustable and determines the depth of the cut. Inv. 1894–171.

19. DOUBLE-FURROW PLOUGH. (Scale 1 : 5.)

The double plough was not extensively used until the beginning of the nineteenth century.

The model represents a double-furrow plough made by Mr. J. P. Fison about 1872. The plough has two parallel iron beams stayed together and each carrying a share, mould-board, and coulters. One of the beams is adjustable sideways so as to vary the pitch of the furrows from 7·5 in. to 10·5 in., or it may be entirely removed. The main beam has long handles at its rear end, by means of which the plough is guided and turned, while its front end carries an adjustable head, having an eye through which passes a draw bar attached to the beams between the two plough frames. The shares and mould-boards are fitted to wrought-iron frames bolted to the beams, and the coulters have round shanks held in sockets and clamped to the beams by eye bolts, their angles being adjusted by set screws bearing on the beams. Skim coulters, whose function is to pare the surface of the land and turn in vegetation to ensure its complete burial, are fitted in front of the main coulters. The front end of the plough is carried by a pair of wheels, one of which runs in the furrow and the other on the land. These are adjustable vertically and regulate the depth of the furrow. A smaller wheel at the rear end, running in the furrow, carries the weight and reduces the draught. The land side of the rear frame is fitted with a plate which bears against the vertical side of the furrow and takes the thrust of the mould-board. The dimensions of the standard slice are 9 in. wide by 6 in. deep, and it is turned through an angle of 135 degrees. Inv. 1906–13.

20. STEAM BALANCE PLOUGH. (Scale 1 : 16.) Presented by Major A. Steinmetz.

This is a model of a multiple plough for one-way work with steam tackle, such as is used in this country at the present day.

A number of plough bodies are mounted on a frame, half of the bodies are built to turn furrow-slices to the right and the other half on the opposite side of the wheels to turn the furrow-slices to the left. The frame is so constructed that when the right-handed bodies are in work the left-handed bodies project forward at an angle of about 40 degrees to the horizontal and so are held clear of the ground.

Two large steam engines are used ; each is equipped with a winding drum and rope and one engine is placed at either end of the piece of ground to be ploughed. The ropes from the winding drums are attached to the rings which can be seen on the under-side of the central wheel carriage of the plough.

The plough is drawn up at one end of the field, while the engine at the other end hauls on its rope. The tension exerted by the rope on the wheel carriage draws the plough-bodies on the opposite side of the wheels down into their work, and the plough is steered by a man in the seat provided. When the plough has been drawn across the field, the engine at the end from which the plough started, and which has been so far paying out rope, pulls on the rope and draws the opposite set of plough-bodies down into their work. The operator on the plough moves across to the other seat and steers the plough back across the field. While this is happening the engine which is not pulling pays out rope and moves along the headland ready for its next pull.
Inv. 1927–826.

21. THE KENT PLOUGH.

This type of plough is found in use in East Kent, the Weald and East Sussex at the present day, and it is constructed so that the furrow slice can be turned either to the right or to the left.

The specimen shown was probably built 70 to 80 years ago, but it does not differ essentially from the modern Kent plough. This kind of plough was in use in Kent

46

certainly 300, probably more than 400, years ago, and it is said to have been imported from the North of France.

A distinctive characteristic of the Kent plough is that it has always been a one-way plough. That is, ploughing is begun at, say the eastern end of the field and the furrow slice is thrown to the right when the plough is moving northwards, and to the left when it is moving southwards. The plough is now set to throw its slice to the right ; to throw the slice to the left the following adjustments are made. The iron wrest below the right-hand mould-board and the iron peg which holds the tail of the wrest out across the furrow are detached and replaced on the left side of the plough. The point of the coulter which is now above the left edge of the share, is moved across to the right edge of the share by adjusting the rod which lies above the beam. This rod lies with its middle part on the left of the coulter, and its two ends on the right of the forward peg and the part of the sheath which projects from the beam. When the slice is to be turned to the left, the middle of the rod will be to the right of the coulter and the ends to the left of the peg and the sheath.

Inv. 1927–972.

22. ONE-WAY PLOUGH. (Replica copied from an original Plough in the Museum, Torquay.)

This type of one-way plough had some popularity on the hillsides of South Devon in the early nineteenth century, but it has been gradually replaced by the balance plough. It was known locally as a " One-way Zuel," and occasionally ploughs built on this principle are still met with in the West.

The heavy wooden share beam, the two wooden sheaths and the wedges to hold the coulters in position, are suggestive of very early days, but the metal coumbs running up from the shares and leading to what were probably metal mould-boards suggest the late eighteenth century.

The mould-boards of the plough are absent, and were probably of metal:

Inv. 1927–997.

23. LOWCOCK'S PLOUGH. Presented by Arthur H. Ogilvie, Esq.

This type of plough was constructed by Messrs. Ransomes, Sims and Jefferies, Ltd., as early as 1850, and was very popular in South Devon in the middle of the nineteenth century. Specimens can still be found in that district.

Comparison of this plough with its wooden counterpart will suggest how the swing mould-boards were arranged on the earlier plough, and comparison with the Howard long-plate plough will draw attention to the narrow share and highly tilted narrow mould-board of the Lowcock plough, features which were common to all ploughs up to the middle of the nineteenth century. Inv. 1928–1200.

24. MODERN ONE-WAY PLOUGH. Lent by Messrs. J. Huxtable and Son.

This plough embodies patents obtained by Mr. John Huxtable in 1877 and subsequently. The type is in general use at the present day in the West Country, particularly in N. Devon and in Somerset.

In this plough a right-handed body and a left-handed body are mounted one above and one below the beam, so that while one is in work the other is carried in an inverted position vertically above it. The frames of the two bodies, hollow beam and the casting to which the draft chain is attached are all in one piece. A bar runs through the hollow beam and carries the handles at its hinder end, thus when the position of the bodies is reversed, the handles are held in position, while the hollow beam rotates longitudinally around the bar. The cross bar which carries the two wheel standards also carries a bevel pinion which meshes with another of the same diameter mounted on the forward end of the bar which runs through the beam. By this means, as the plough bodies are rotating on the axis of the beam, the wheel standards are also rotated through 180 deg. and are brought into the required new position. The plough bodies and wheels are held in the correct position relative to the handles by a catch, located where the handles and the hollow beam meet.

SCOTCH PLOUGHS

25. SMALL'S WOODEN PLOUGH. (Scale 1 : 4.)

This model, an exact copy of an earlier model in the possession of the Royal Scottish Museum, Edinburgh, represents a plough introduced into Scotland by James Small, about 1760.

There are statements to the effect that others brought this plough to Scotland prior to Small, but all the available evidence points to the fact that the credit for its introduction lies with him. Such a plough as this is capable of being drawn effectively by only two horses, and ploughs which it displaced were much heavier and clumsier, resembling the old Saxon plough and needing the power of six or eight oxen.

Inv. 1928–1369.

26. WOODEN PLOUGH. (Scale 1 : 4.) Lent by The Royal Scottish Museum.

This type of plough is a modification of the Rotherham pattern, and is typical of the ploughs which were becoming popular in Scotland about 1780.

The mould-board is remarkable in that it is made of iron, but its shape appears to be rather more primitive than that of the Rotherham ; unfortunately, the share is missing. The curve of the beam corresponds to the accepted form of the Rotherham beam, and the way in which the land handle runs on from the end of the beam to join the sheath is a feature which indicates very clearly the relationship to the Yorkshire plough.

Inv. 1928–345.

27. SMALL'S IRON PLOUGH. (Scale 1 : 4.)

This model, an exact copy of a model in the possession of the Royal Scottish Museum, Edinburgh, represents one of the earliest iron ploughs made by Small.

It is interesting to note that in the iron plough Small has reverted to the older quadrilateral frame, probably in order to obtain the required length for the sole, and then to be able to attach it to the beam with a frame which combines lightness with strength. The hake is also interesting in that it is an older form than the Suffolk hake, which he has used on his wooden plough. In this model the form of the share and of the mould-board show distinct improvement on those parts of the wooden plough.

Inv. 1929–779.

28. EAST LOTHIAN PLOUGH. (Scale 1 : 8.) Lent by The Royal Scottish Museum.

The plough represented indicates the influence of the Rotherham plough on Scottish plough design.

The elliptical breast and the Suffolk pattern of hake (draught iron) suggest that the maker followed not the original Rotherham plough of 1730, but rather the Rotherham as it was improved and used in Norfolk about 1770.

Inv. 1928–487.

29. EAST LOTHIAN PLOUGH. (Scale 1 : 4.) Lent by The Royal Scottish Museum.

This type of plough was intended for use with a single horse instead of the usual team of two or three horses, and is similar in outline to the earlier Lothian plough, but is more lightly built.

Inv. 1928–399.

30. EAST LOTHIAN PLOUGH. (Scale 1 : 4.) Lent by The Royal Scottish Museum.

This early nineteenth-century model is more heavily built than that intended for single-horse work, and the twist of its mould-board is more marked, but like most of the contemporary E. Lothian ploughs, it has the long thin share and curved breast of the mediæval plough. The fact that a long share gives a slower rise to the breast, and an easier twist to the mould-board, would militate against any tendency to shorten the share where the plough is to be used for autumn work. Moreover, the long breast and share would allow the plough to penetrate more easily and when correctly set would tend towards lighter draught.

Inv. 1928–451.

31. CURRIE PLOUGH. (Scale 1 : 4.) Lent by The Royal Scottish Museum.

The important feature of this East Lothian type of early nineteenth century plough, which the maker sought to stress, is the short mould-board.

Although the plate is shorter than usual, the twist is more marked than in other Lothian ploughs. This fact, coupled with the broad flat share and the quick rise of the breast, suggests that the plough was intended particularly for spring work. If such is the case, the plough is an interesting instance of an early attempt to differentiate between the digging plough for spring work and the long-plate plough for autumn work. The draught rod and the unusual hake, with its great capacity for adjustment, are notable features.

Inv. 1928–452.

32. SUBSOIL PLOUGHS. SMITH'S PLOUGH AND SLIGHT'S PLOUGH. (Scale 1 : 4.) Lent by The Royal Scottish Museum.

Sub-soiling is the operation of stirring the layer of soil which lies below that which is treated by the plough and the other normal tillage implements.

These two ploughs are typical of the implements that were used for this purpose.

48

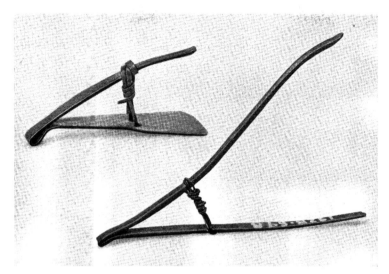

Egyptian Digging Implements (Cat. No. 33).

Russian Plough (Cat. No. 39).

Slight's claim that his is a two-horse implement may be somewhat optimistic as sub-soiling is notoriously hard work, though this implement would work in the furrow following a common plough.

At the present day the usual practice is to remove, on a two or three furrow tractor plough, one of the plough bodies, preferably the leading one, and to fix in its place an object resembling a stoutly-made cultivator tine, or to make use of steam tackle.

Inv. 1928–401, 343.

FOREIGN PLOUGHS

33. EGYPTIAN DIGGING IMPLEMENTS. (Scale about 1 : 16.) Presented by A. Lucas, Esq.

These are two copper models of digging implements taken from the Theban Necropolis. They indicate the kind of tools used by the Egyptians in the XVIIIth Dynasty, about 2000 to 1500 B.C.

Certain of the illustrations which show phases of life in ancient Egypt, depict men digging the soil and using an implement similar to the larger of these two models. The end of the longer curved handle was held in both hands and the action was exactly similar to that used with the modern pick-axe. The design of the Syrian plough and the Russian plough appears to be closely related to that of these implements.

Inv. 1928–648.

34. ANCIENT EGYPTIAN HOE. Presented by the Egypt Exploration Society.

This implement was found near Tel-el-Amarna, Upper Egypt, and is considered to have been in use about 1500 B.C.

The hoe is made of bronze, and it was probably intended to have a straight wooden handle and to be used as a push hoe.

The fact that it seems to have been subjected to wear more on the side of the edge which lies away from the handle suggests the push hoe or the plough. The edge is worn more at one corner than at the other ; this and other slight indications suggest that it was part of a push hoe, and that it was used with the join of the socket upper-most.

Inv. 1929–660.

35. COLLECTION OF MODELS OF PRIMITIVE PLOUGHS. Lent by Major Steinmetz (*except where otherwise indicated*).

The specimens are all models of primitive hoe ploughs. They indicate the kind of plough which was in use before the Christian Era, and they show how the early types linger in backward countries. None of these ploughs turns a furrow slice to one side, but all are designed to stir the soil in a manner resembling the modern cultivator tine.

The two Egyptian implements shown were hand implements which were used for digging in a manner similar to the pick-axe, but it is easy to see their close connection with the earliest type of plough. The Syrian plough is a model reconstructed from an illustration, and shows what is perhaps the earliest known form of plough. The two Roman ploughs are not typical, but they have interesting features. The wheel plough is really only the coulter ; the share and remaining parts were drawn behind it by a separate team of oxen, a practice which persisted in Spain in the eighteenth century. The other wheel-less Roman plough is an early attempt at a ridging plough ; the soil was split by the share and thrown to both sides by the wrests and the two upright pegs. The Russian plough resembles the early Egyptian ploughs except for its double share.

Models of ploughs from Egypt, North Africa, India, East Indies and Mexico are also shown. The similarity between all these types and the Syrian is striking.

The Chinese ploughs with their rectangular frame are unique in this collection, but it is curious that their form approximates to that of the Teutonic plough.

The Steinmetz collection (all to a scale of 1 : 16) includes the following types : Early wheel, Asia Minor, Modern Egyptian, Moroccan, Senussi, Indian, East Indian, Mexican, Ancient Chinese, and Modern Chinese.   Inv. 1926–816 to 822, 824 to 827.

36. SIAMESE PLOUGHS. (Scale 1 : 8.) Lent by L. H. Pritchard, Esq.

These models represent the primitive form of plough used in the rice fields of Siam. They are made entirely of wood and consist of a long curved pole to which a yoke is attached, and a cross piece mortised on at the rear end, the lower end of which carries the ploughshare, while the upper end serves as a guiding handle. The share is formed with an upward projection having a curved face which slightly turns the

D

shallow furrow made. The ploughs are drawn by a pair of water-buffaloes, the yoke bar resting on their necks and being secured by a cord passed underneath.

Inv. 1908–14.

37. MOROCCO PLOUGH. Lent by Major A. S. B. Steinmetz.

This type of plough is made by the natives of Morocco, where it is in use at the present time.

The natives make this plough in three sizes for drawing by donkeys, camels, or buffaloes. The body of the implement is made from the heart of a tree, the handle being roughly spliced on and the beam morticed through ; a heavy pointed iron share is fixed to the body. Inv. 1926–850.

38. SENUSSI PLOUGH. Lent by Major A. S. B. Steinmetz.

This plough was seen working on the borders of the Libyan Desert in 1915.

The bottom of this plough was made from the handrail of a shipwrecked vessel, and when working the plough was pulled by a camel. Inv. 1926–849.

39. RUSSIAN PLOUGH. (Scale 1 : 4.) Lent by The University of Edinburgh.

This is a model of a plough which is still in use in Russia, Finland, Esthonia, etc. The model was presented to the Highland and Agricultural Society of Scotland in the middle of the nineteenth century.

It is a type which is peculiar to Eastern Europe, and has some similarity to the early Egyptian pattern (No. 33). The plough represented by this model was apparently used as a one-way plough. It was drawn by a man or a single horse, and by tilting it to the left the soil could be turned to the left or *vice versâ*.

Inv. 1928–634.

40. SIAMESE PLOUGH. (Scale 1 : 10.) Presented by J. W. Hinchley, Esq.

This is a model of a native Siamese plough, which is constructed of wood with a metal shoe or blade. A yoke is attached to the long curved pole. The plough is drawn by a pair of water buffaloes. Inv. 1914–131.

41. COLLECTION OF MODELS OF MODERN FOREIGN PLOUGHS. (Scale 1 : 16.) Lent by Major Steinmetz.

One feature which is common to all the ploughs in this collection is the American pattern share and mould-board. This form of body is used by American makers and by English manufacturers for the Colonial markets. It could be used for spring work in this country, but it is never likely to compete seriously with the English plough for autumn work.

The collection includes models of the following types : Siamese, Cingalese, Asia Minor, Indian, Siberian, South African, American, Bulgarian, German, Russian, African elephant, Italian turn-wrest, and Spanish turn-wrest.

Inv. 1926–829, 830, 833 to 843.

42. EGYPTIAN PLOUGH. (Scale 1 : 4.) Presented by Fouad Bey Abaza.

This model is an example of a modern Egyptian plough, but it differs very little in outline from certain ploughs which are included in pictures of Ancient Egyptian Agriculture.

As the model is arranged, it is ready to be used for cutting furrows, throwing the soil to both sides. When it is needed for ordinary ploughing a block between the beam and the share beam is removed. Inv. 1926–1062.

43. DISC PLOUGH. (Scale 1 : 16.) Lent by Major Steinmetz.

This plough was first introduced for the sticky soils of Mexico, and is now also in use in Australia, South Africa, and South America. As the discs revolve, the scrapers remove any adhering soil from them. Inv. 1926–830.

CULTIVATORS AND HARROWS

44. WILKIE'S GRUBBER. (Scale 1 : 4.) Lent by The Royal Scottish Museum.

This type of cultivator was designed about the middle of the nineteenth century.

The tines are raised by links which are actuated by the handles. The points of the tines are fitted with a broad flat share which equips the cultivator particularly for work on a stubble after harvest. The whole width of the cultivator is covered at one point or another by the cutting edge of a tine, and while the soil is loosened, annual weeds are cut off and turned out on to the surface. By this means weeds that are growing are killed, and weed seeds in the soil are encouraged to germinate ; then two or three weeks later, when the latter are in a tender stage of growth, they are also killed by a stroke with the harrows. Inv. 1928–402.

45. FINLAYSON'S CULTIVATOR. (Scale 1 : 4.) Lent by The Royal Scottish Museum.

This implement was invented by Finlayson early in the nineteenth century, and it is stated in the " Gardener's Magazine " that it achieved some degree of fame by being used for " Breaking up Hyde Park, London," in 1826.

The earlier cultivators carried seven or nine perpendicular rigid tines which forced their way to their working depth by virtue of the weight of the implement. Finlayson's cultivator indicates progress in the forward slope of the tines, which find their way down to their working depth almost regardless of the implement's weight.

Clods and weeds tend to slide up the tine and arrive at the soil surface, where the former can be dealt with by harrows and roller and the latter killed by exposure.

Inv. 1928–346.

46. BINNIE'S GRUBBER. (Scale 1 : 4.) Lent by The Royal Scottish Museum.

This cultivator was introduced into Scotland by Binnie about the middle of the nineteenth century.

The tines do not pitch forward so sharply as in Finlayson's implement, but the lifting mechanism is an attempt at something more efficient. The hind wheels are mounted on a separate carriage, which is connected directly at two points to the hind bar of the frame and by levers to the apex of the frame and the front wheel. When the handles of the wheel carriage are depressed, the back bar of the frame is raised as well as the forward point of the frame, so that all tines are raised on an even keel. When the tines are in work the hind wheels are held up by the anchor at the top of the implement, and they are prevented from drawing in to too great a depth by the adjustment of a sliding collar on the bar which runs back horizontally from the front wheel standard. Inv. 1928–349.

47. KIRKWOOD'S GRUBBER. (Scale 1 : 4.) Lent by The Royal Scottish Museum.

This cultivator is very similar to Binnie's grubber except that the depth of working is controlled by the screws on the upper horizontal bar. Inv. 1928–454.

48. MARTIN'S CULTIVATOR. (Scale 1 : 4.) Presented by Martin's Cultivator Company.

This form of cultivator, which was patented by Mr. W. C. Martin in 1898, is used for breaking up and pulverizing the soil, preparing seed beds, and other similar operations. The size represented has seven tines and works a width of 5·25 feet to a depth of from 1 to 8 inches.

The implement consists of a frame at the back of which is supported a transverse shaft having at its ends crank arms with pins carrying the wheels. At intervals along this shaft are placed blocks clamped to a flat bar behind them. The tines are formed of rigid curved bars, to the lower ends of which are bolted reversible shares arranged in two rows, whose points cut the soil tangentially. The front end of each tine passes freely through slots in the lower part of one of the blocks and is attached by a pin to one end of a bent spring whose other end is clamped to the block and cross bar. By this arrangement the tine has a vibratory motion which is limited by the slots so that it cannot pull back too far or rise out of the ground. To the wheel shaft is attached a lever working round a quadrant fixed to the tine bar, and by means of this the wheel cranks may be set at any angle, so regulating the depth of cut and keeping all the shares at the same level. When turning, the tines are lifted clear of the ground by another lever which tilts up the tine bar. The forward end of the frame is carried by a small wheel with a swivel mounting and an arrangement for adjusting the height. A seat is provided for the attendant. Inv. 1905–4.

49. SHETLAND HARROW. Lent by The University of Edinburgh.

This type of harrow has been in use in the Shetland Islands up to a fairly recent date. A similar kind of harrow is shown on the Bayeux tapestry (11th century c.) and the type was probably used by the Saxons in this country considerably earlier.

Harrows made in this rectangular form are always drawn from one corner so that the ground is stirred as evenly as possible over the whole breadth covered by the implement. Inv. 1928–627.

51

50. SIAMESE HARROW. (Scale 1 : 10.) Presented by J. W. Hinchley, Esq.

This is a model of the harrow generally used in Eastern countries. A yoke is attached and the harrow is drawn by oxen or buffaloes. Inv. 1914–132.

51. SIAMESE HARROWS. (Scale 1 : 8.) Lent by L. H. Pritchard, Esq.

These models represent the form of harrow used in Siam and other parts of the East. They are simply large wooden rakes provided with a guiding handle and drawn by buffaloes yoked to the pole. Inv. 1908–15.

52. HARROW. (Scale 1 : 4.) Lent by The Royal Scottish Museum.

This kind of harrow was used in Scotland in the nineteenth century. It was also used with a pair of wheels fitted one at each end of the hinder bar and a third wheel in the centre in front, when it functioned as a cultivator.

In spite of its superficial similarity to the Siamese harrow the type represented a more efficient implement. The forward pitch of the tines would draw it down into its work, but a tendency to bury itself would be counteracted by the upward pull of the draught chains, while control of the depth is effected by the regulation of pressure on the handles. Inv. 1928–555.

53. WOODEN HARROWS. (Scale 1 : 4.) Lent by The Royal Scottish Museum.

These models represent the lightest implements in the range of light, medium and heavy harrows which are used on a farm.

The use of timber for the frame of a harrow is no safe indication of its date, as at the present day on light soils timber-framed harrows are used widely.

The jointed harrows are so constructed in order that inequalities of surface shall not upset the even treatment of the soil. Inv. 1928–407–8.

54. IRON HARROWS. (Scale 1 : 4.) Lent by The Royal Scottish Museum.

These models represent medium or heavy harrows and with the models of wooden harrows indicate the form which was popular about 1830 and onwards before the introduction of the zig-zag harrow. Inv. 1928–409.

55. REVOLVING HARROWS. (Scale 1 : 4.) Lent by The Royal Scottish Museum.

These implements have little pulverizing effect, but are very effective in separating weeds such as couch from the soil. The design was taken from an implement which had been in use for many years and still is popular in Norway.

The smaller model is an example of the lighter type of harrow that was made, and the larger represents some refinement of the principles involved. In the latter the axles are set at an angle to the line of advance, so increasing the stirring effect on the soil and enabling the implement to search more thoroughly. After the couch is brought to the surface it is caught up by the following rake. Inv. 1928–350, 382.

56. LAND ROLLER. (Scale 1 : 4.) Lent by The University of Edinburgh.

This model of a cast-iron smooth roller is typical of the kind of roller which has been in use over most of the United Kingdom for over a hundred years.

The full-sized implement would be a heavy, strongly built, and therefore useful specimen needing two good horses to pull it.

A wagon for transporting the roller from field to field is also exhibited to the same scale. Inv. 1928–617–8.

57. CLOD-CRUSHER. (Scale 1 : 6.) Presented by W. Crosskill, Esq.

This model represents the clod-crusher patented by Crosskill in 1841. It consists of a number of cast-iron discs threaded loosely on a horizontal axle-tree. The edge of each disc has projecting V-shaped teeth, and also lozenge-shaped teeth on the sides nearly radial in direction. Two smooth wheels of large diameter are placed on the extremities of the axle when the implement has to travel for some distance over roads. Inv. 1857–113.

58. CLOD-CRUSHER. (Scale 1 : 6.) Lent by Messrs. Newton and Sons.

This is a model of W. C. Cambridge's clod-crusher, patented in 1844. It has a number of cast-iron discs placed side by side on a horizontal axis, but they are alternately toothed and plain on the edges. The toothed discs are greater in diameter than the others, but, having a larger hole in the centre, all bear equally on the soil ; the sliding movement so resulting prevents clogging in moist ground. Inv. 1862–15.

59. CLOD-CRUSHER. (Scale 1 : 4.) Lent by Messrs. R. Garrett and Sons.

This model was shown at the 1851 Exhibition, but the arrangement was invented prior to 1845. It is a form of clod-crusher in which, for moving the machine over hard roads, two travelling wheels carried on independent overhanging axle-trees are introduced. By means of screws and suitable guides the axles can be quickly elevated or depressed, so that when on the fields the travelling wheels can be lifted well above the soil. Inv. 1894–167.

60. SIMAR ROTO-TILLER. Presented by Messrs. Piccard Pictet and Co., Ltd.

The Simar Roto-tiller was designed in 1910 and introduced into this country in 1920. Its purpose is to break up the soil, and produce a tilth in one operation, top-growth or manure being mixed with the soil in the process.

The example shown was used very extensively for agricultural research work at the Rothamstead Experimental Station, Harpenden, but the design has been superseded by a more modern form.

A small, single-cylinder, two-stroke engine which will burn petrol, paraffin or alcohol is mounted on a pair of wheels. It is provided with two gears, the top one giving a running speed of two miles an hour, the bottom one 0·75 of a mile an hour. The wheels are driven by a worm shaft which is extended backwards and which drives, by means of a bevel pinion and crown wheel, the miller shaft which lies parallel to and behind the axle of the ground wheels.

The miller shaft normally runs at about 150 r.p.m., and can operate independently of the ground wheels. The sleeves are mounted at either end of the miller shaft, and each carries six specially designed coiled springs. A semi-circular tine of hardened, tempered steel is fitted to the extremity of each of the twelve springs. The whole miller is enclosed under a hood to prevent the scattering of soil and stones, the hood being hinged to facilitate inspection.

The rotating tines churn up the soil, over a width of 20 inches and to any depth required between 2 and 10 inches. Narrower millers can be obtained or on modern machines the two sleeves can be fitted a short distance apart to work on either side of a row of plants. Specially shaped tines can be fitted for other types of work such as scarifying the surface of arable land between the rows of a crop or for aerating pasture. The machine can be fitted with a mower attachment or it can be adapted for belt work or for drawing a ridging plough, a seed-drill, a lawn-mower or a trailer.

Inv. 1929–371

61. MODERN SIMAR ROTO-TILLER. Lent by Messrs. Piccard Pictet and Co., Ltd.

This is a modern type of rotary tiller, and comparison with the older machine (No. 60) will show what progress has been made in the details of construction.

The main principles of the machine remain the same. The engine has been made more than twice as powerful, and there are slight differences in the lay-out of the engine, including an improved air cleaner which is placed well forward. The wheels are larger, giving better clearance under the axle and a better grip of the soil. The improved design of the handles gives better control, and the adjustments are more easily operated. The mill is more thoroughly covered with a hood which is made in one piece. Two sleeves are fitted on the mill shaft and the tines are carried on the sleeves so that the machine can be worked astride a row of plants.

Inv. 1929–541.

62. TINES FOR USE WITH THE SIMAR ROTO-TILLER. Lent by Messrs. Piccard Pictet and Co., Ltd.

These are specimens of the different tines which can be used for various purposes with the Simar Roto-tiller.

The normal tine for digging soil to any depth beyond four inches is the narrow chisel-shaped type. The chisel-shaped tines with broader edges are for stirring the top two to four inches of soil between the rows of a growing crop in summer. The knife-edged tines are used for cutting turf and in opening up and aerating matted grass land. Inv. 1929–942.

SEED DRILLS AND MANURE
DISTRIBUTORS

63. HAND DIBBLER. Lent by The Royal Scottish Museum.

This implement is used for making holes in the soil for planting potatoes or transplanting cabbages. Inv. 1928–524.

64. MECHANICAL DIBBLER. (Scale 1 : 3.) Lent by The Royal Scottish Museum.

This model was submitted to the Highland Agricultural Society in 1835, and represents a machine intended to dibble wheat rapidly.

The machine is intended to be held upright by the top bar, and the points are placed in the correct position on the soil. The operator then presses downwards, driving the points into the soil, and actuating the mechanism which causes a seed to be tipped down each tube. Inv. 1928–356.

65. BEAN DRILL. (Scale 1 : 4.) Lent by The Royal Scottish Museum.

This early nineteenth century bean drill was intended to be driven by hand along the furrow after the plough ; as it travelled it dropped a thin line of beans in the furrow, which were covered by the next furrow-slice turned by the plough.

In this machine, as in the earlier types, a crank on the ground wheel drives by means of a connecting rod, a short shaft which runs through the bottom of the hopper. A grooved roller carried on this shaft, rotates with it and passes the beans down so that they fall on to the soil. Inv. 1928–455.

66. BEAN DRILL. (Scale 1 : 4.) Lent by The Royal Scottish Museum.

This bean planting attachment for the plough was introduced before 1850, and the idea is still carried out in the form of smaller and more compact machines for attachment to the modern plough.

The machine was slung between the handles of a plough, and the two hooks at the back of the hopper were attached to the handle stay of the plough. The wheel runs along in the furrow, and drives a grooved roller which passes the beans out into the furrow behind the plough. Inv. 1928–412.

67. MORTON'S GRAIN DRILL. (Scale 1 : 4.) Lent by The Royal Scottish Museum.

This machine, invented by Morton in 1828, is designed to sow wheat, barley or oats in rows about 4 to 6 inches apart.

The machine is designed to deposit the seed in five drills at a time. In each drill a small furrow is cut by a cultivator tine and a line of seed is allowed to fall into the furrow. The large wheel drives, through a universal joint, a shaft on which are mounted five grooved rollers, which pass the seed from the bottom of the seed hopper to the ground. In this way the amount of seed deposited along the drill will be constant for any speed at which the drill may move. The rate of sowing is adjusted by slides which lie just above the rollers and the depth at which it is deposited in the soil is altered by varying the adjustment of the two travelling wheels. The two tines which protrude on either side of the drill are for marking the unsown land so that the driver has a guide as he returns with the drill across the field. Inv. 1928–410.

68. CORN DRILL. (Scale 1 : 4.) Lent by Messrs. R. Garrett and Sons.

This is a model shown at the 1851 Exhibition, and represents a machine for drilling corn into the soil in parallel rows, and delivering a uniform amount per yard irrespective of the speed at which the machine moves. Details of this machine were patented in 1842 and 1846, but the general arrangement is substantially that still followed. The body of the machine is carried on two large driving wheels, and an independent fore-carriage is employed by means of which the attendant may accurately steer the machine and thus avoid irregularity in the sowing—a matter of considerable importance if a horse hoe or other implement is afterwards to be employed between the rows. The seed is carried in a hopper at the top of the casing, from which it

54

passes through five adjustable doors into the trough. Above this trough is a horizontal shaft, driven by spur gearing from one of the travelling wheels, and having on it five discs each provided with metal cups which, as the discs revolve, dip down into the seed and lift it up, in the same manner as the old Persian bucket-wheels raised water. On both sides of each disc is a spout into which the elevated seed drops, and by a series of nested funnels is led down to a groove formed in the soil by the drill coulter—the model having ten of these. Each drill is carried on a lever so that it will rise independently and not be damaged should it meet with any solid obstruction, while a windlass at the back of the machine raises all the drills simultaneously from the ground, so that the depth of the sowing can be readily altered. The quantity sown is varied by the use of change-wheels, while to correct the level of the hopper when on sloping ground a screw adjustment is provided. When travelling only, the feed mechanism can be thrown out of gear by a lever which lifts the overhanging toothed wheel. Inv. 1894–162,

9. DUDGEON'S ROOT DRILL. (Scale 1 : 4.) Lent by The Royal Scottish Museum.

This drill is designed to sow turnip seed on soil which has been thrown into ridges with the double mould-board plough, also it is constructed so that artificial manure can be deposited in the row with the seed. It was introduced into Scotland between 1830 and 1840.

The manure is placed in the large hopper and is passed down to the soil by a grooved roller. The turnip seed is placed in the smaller barrel-shaped hopper and falls through small holes in the equator of the barrel as it rotates. The soil is first opened by the hoe coulter. The manure is then deposited and is partially covered with soil by a roller. A smaller hoe coulter makes way for the seed which in its turn is covered by the heavy roller which follows. By this means contact between the manure and the seed is to a great extent avoided, an arrangement which is an advantage when it is necessary to use some artificial manures which may do harm to the seedling. Inv. 1928–352.

70. NICOL'S ROOT DRILL. (Scale 1 : 4.) Lent by The Royal Scottish Museum.

This is a model of a mid-nineteenth century drill designed to deposit manure and seed on the ridge in a similar manner to Dudgeon's drill.

The delivery mechanism for the manure and for the seed is also similar. In this case manure and seed are allowed to run together into a shallow furrow cut by the hoe coulter and they are covered with soil by a tine and a roller. Inv. 1928–353.

71. LISTON'S ROOT DRILL. (Scale 1 : 4.) Lent by The Royal Scottish Museum.

This mid-nineteenth century drill is designed to sow manure and seed on the ridge, and it resembles Nicol's and Dudgeon's drills in its delivery mechanism.

The row is opened by a conical roller, the manure and seed are fed into the furrow by a mixer and are then covered by the roller. The mixer takes the form of a grooved roller mounted on each delivery tube, close down to the soil. A small quantity of manure and seed is collected in each groove and is carried round and allowed to fall into the furrow as the roller rotates. This mechanism also has the effect of dropping the little parcels of seed and manure at regular intervals along the row, and so effecting some further economy of seed and minimizing the work of thinning the crop after the plants are showing above ground. Inv. 1928–354.

72. ROOT DRILL. (Scale 1 : 4.) Lent by The Royal Scottish Museum.

This is a model of a mid-nineteenth century drill for sowing turnips either on the ridge or on the flat.

It is very noticeable how much lighter and less complicated the drill becomes when it is built for sowing seed alone. It is equipped with the usual barrel delivery and the hoe coulter and roller. Inv. 1928–411.

73. EAST LOTHIAN ROOT DRILL. (Scale 1 : 4.) Lent by the Royal Scottish Museum.

This is a model of a mid-nineteenth century drill for sowing turnips on the ridge.

It has the hoe coulter and roller and the barrel delivery, which at that time were the standard design, but it also has the following features which had not appeared previously. The driving rollers are concave and travel ahead of the coulters, compressing the ridge before the seed is sown. The two seed barrels are driven independently each by a separate strap from its own roller, thus either can be thrown out of action without affecting the other. Inv. 1928–456.

74. TURNIP DRILL. (Scale 1 : 4.) Lent by Messrs. J. C. and T. Yates.

This is a model of a machine for sowing turnips and similar roots. The seeds are drilled on ridges and the machine is arranged for drilling two ridges at once. It has two light travelling wheels, from one of which the mechanism for the ordinary seed-lifting cups is driven ; this and the seed being confined in a central box. From the box the lifted seed is conducted to the two coulters, which by attached levers are readily under the control of the attendant. In advance of each coulter runs a wide concave metal roller, which by its pressure prepares the ridges into which the seed is drilled. Behind the coulters follows a timber roller which, pressing on the ridges, consolidates the earth over the seed. Inv. 1894–172.

75. GENERAL SEED DRILL. (Scale 1 : 4.) Lent by Messrs. R. Garrett and Sons.

This model was shown at the 1851 Exhibition, but the arrangement was patented in 1844. It is a seed drill for general purposes, arranged to drill in seed and any suitable manure at the same time. The manure and the seed are contained in two separate chambers at the top of the machine. The seed is dealt with in the same way as with an ordinary corn drill, but beneath the seed mechanism is arranged a shaft with a number of arms terminating in cups, by which the powdered manure is thrown into conical hoppers that deliver it into the conductors conveying the seed to the coulters. Various arrangements of the coulters can be made so as to give several different distributions of the seed and manure. The seed mechanism is worked from one of the driving wheels, and the manure distribution from the other, but either can be thrown out of gear by the hand levers provided. Inv. 1894–164.

76. INDIAN SEED DRILL. (Scale 1 : 2.) Lent by The Royal Scottish Museum.

This model represents the type of drill which is made and used by the Indian peasant.

The seed is placed in the saucer-shaped depression in the teak spreader. The cone in the bottom of the saucer is perforated at its base by twelve holes through which the seeds pass into the twelve bamboos and so to the soil. The short bamboos below the horizontal beam act as delivery tubes and also as hoe coulters. There is no mechanism for seed delivery and no adjustment for rate of sowing. Depth of sowing is controlled within limits by the pressure of the man's hand on the guiding handle behind the drill. Inv. 1928–515.

77. CARROT DRILL. (Scale 1 : 4.) Lent by The Royal Scottish Museum.

This machine for drilling carrots was invented by Daniel M'Naughton, an Ayrshire farmer, in 1833.

It is similar in general appearance and lay-out to the contemporary turnip drill, but differs in its delivery mechanism. The transverse shaft which is driven by a belt from the leading roller carries two pulleys armed with spikes, which when they rotate scrape the seed from the bottom of the hopper, and pass it into the delivery tubes. The shaft also carries two pulleys which drive two other pulleys just above them ; these actuate spiked rollers mounted inside the bins for the purpose of stirring the seed. The drill can be adapted for turnips by removing the bins and the transverse shaft and fitting another transverse shaft on which are mounted two of the ordinary seed-barrels. For sowing onions a shaft carrying five smaller barrels is mounted, and the five-coultered bar is attached instead of the three-coultered bar, the necessary extra two delivery tubes being also added. Inv. 1928–525.

78. GRASS SEED BROADCASTER. (Scale 1 : 4.) Lent by The Royal Scottish Museum.

This model was awarded a premium by the Highland Society in 1836. The machine allows grass seed to fall uniformly on the soil and by means of the roller covers the seed with a thin layer of soil.

The grass seed is placed in the hopper and is scooped out regularly at the bottom by toothed wooden rollers mounted on a transverse shaft, the shaft being driven from the large following roller. Although the seed is delivered through regularly spaced holes underneath the box it has sufficiently far to fall to ensure that it spreads fairly evenly over the soil.

An attachment for harrows is provided, the object being to ensure the burial of the seed and to loosen the surface of the soil after the roller. Inv. 1928–500.

56

79. ARTIFICIAL MANURE DRILL. (Scale 1 : 4.) Lent by The Royal Scottish Museum.

This machine, introduced in 1828, is for distributing such artificial manures as bone dust or soot on the rows of plants. It usually follows the seed drill.

Running along the bottom of the hopper is a shaft which is parallel to the axle, and is driven by it through gear wheels. Mounted on the shaft are three toothed wheels which, in rotating, scrape manure into the three holes leading to the delivery tubes. No coulters, as in the case of seed drills, are provided, as it is not so important to have manure buried promptly. It is sufficient if it is allowed to fall on the surface and be mixed with the soil by the harrows which follow. Inv. 1928–526.

80. SOOT DISTRIBUTOR. (Scale 1 : 4.) Lent by The Royal Scottish Museum.

This distributor was invented by Alexander Main, in 1839.

It was intended for the broadcast distribution of soot on growing crops, and it is interesting in that its delivery mechanism has a close relationship to a certain type of modern manure distributor. The axle carries the two ground-wheels, a long fluted wooden roller which fits into the bottom of the hopper, and a gear-wheel which lies between the left wheel and the roller. The wheels drive the axle, and the roller and gear-wheel turn with it. The gear-wheel drives through an idler, a pinion which lies above it, and is carried on a shaft which runs through the upper part of the hopper. Fixed to this shaft, inside the hopper, is a cylinder of sheet iron, perforated over its whole surface with 0·5 in. holes. Trap doors in the cylinder admit of its being filled with soot ; as the machine moves forward, the rotation of the cylinder allows soot to fall into the hopper and the flutings in the rotating roller below carry the soot out of the hopper and allow it to fall to the ground. A brush is applied along the length of the roller to regulate by its pressure, the rate of sowing of the soot. Foreign matter, such as stones and mortar, is held up inside the cylinder and emptied away before refilling. Inv. 1928–358.

81. LIQUID MANURE DISTRIBUTOR. (Scale 1 : 4.) Lent by The Royal Scottish Museum.

This is a nineteenth century model of a machine for distributing liquid manure over arable or grassland.

The hopper at the top of the barrel carries a strainer and the liquid manure runs by gravity through the tap at the back and into the distributing pipe, finding its way out through the many small holes in the pipe. It was realized quite early that the liquid part of the manure contained highly valuable plant food. Where possible straw was used in plenty to soak it up and avoid waste, but with the march of progress drains were led from the cow-sheds and stables to a single tank, where the liquid could collect and periodically be pumped up into the distributor and taken back on to the land. Inv. 1928–375

82. LIQUID MANURE DRILL. (Scale 1 : 4.)

This is a form of drill for sowing the seed of turnips or other root crops, along with water or liquid manure : it was patented by Mr. T. Chandler in 1847, and has been largely used in the Southern Counties of England, where such crops are sown on flat ground which is frequently dry.

The drill represented has two travelling wheels, sows three rows, and is drawn by one horse. A large wooden trough, situated over the axle, carries the manure, and above it, at the back, is mounted the seed-box, whence the seed is lifted by cups on a shaft driven through a claw clutch and spur gearing from one of the wheels. To economise seed, there are only four cups to each row, and the speed of the shaft is such that the seed is dropped at intervals, and not continuously. The seed falls through nests of funnels to the coulters, the latter being carried on levers, the height and pitch of which can be varied.

The manure trough is mounted on trunnions, and one end can be elevated or depressed, so as to keep the axis level ; it can also be levelled in the fore and aft direction, both these adjustments being controlled from the rear of the machine. The manure is lifted by three buckets, driven by gearing, and is emptied through adjustable openings into a second set of nested funnels, which deliver it into the lower seed funnels. Inv. 1911–59.

83. LIQUID MANURE DRILL. (Scale 1 : 4.)

This represents a combined seed and liquid manure drill, similar to the Chandler drill, but with modifications in construction, and in the method of distributing the manure, patented by Mr. R. Reeves in 1854.

57

The drill has two travelling wheels, and sows three rows ; the manure trough is fixed to the frame, and the seed-box is bolted to the back of it.   The seed is delivered by the cups into flexible leather pipes leading to metal tubes fixed behind the coulters. The manure passes through perforated plates in the bottom of the trough to pipes leading backwards to the coulter tubes ; sliding valves beneath the perforated plates regulate the feed.   The trough contains a revolving stirrer, and also spring blades, which scrape over the perforated plates to prevent them choking.   The pitch and also the height of the coulter levers can be adjusted ; a clutch is provided for disconnecting the seed drill from the driving gearing.   Inv. 1911–60.

RIDGING PLOUGHS AND HORSE-HOES

84. RIDGING PLOUGH. (Scale 1 : 4.) Lent by The Royal Scottish Museum.

This is a model of the kind of plough which is used for throwing the soil into ridges.

The model differs from the modern ridging plough in possessing a coulter, but in other respects it is practically up-to-date. The mould-boards are capable of being adjusted to throw a wide or a narrow furrow ; unfortunately the share is absent.

Inv. 1928–488.

85. COMBINED HORSE-HOE AND RIDGING PLOUGH. (Scale 1 : 4.) Lent by The Royal Scottish Museum.

The implement represented, introduced in 1831, is intended to combine the functions of ridging plough and horse-hoe.

In its present form it acts as a horse-hoe ; the share of the ridger performs the functions of the " A " tine, and two " L " tines mounted on a frame capable of adjustment for width, complete the equipment necessary for a hoe. When the implement is required for ridging, the two " L " tines are removed and two mould-boards are attached, one on either side of the frame. A modern implement somewhat similar to this is also capable of being used as a sub-soiler ; the mould-boards or " L " tines are removed, and it is drawn along the furrow after the plough so that it stirs the surface of the subsoil. It is possible that this implement may also have served this purpose, though it is certain that the provision of a head-wheel would have made its use as a sub-soiler very much more easy. Inv. 1928–351.

86. EXPANDING HORSE-HOE. (Scale 1 : 4.) Lent by The Royal Scottish Museum.

This type of horse-hoe was introduced about the year 1831, and shows improvement in the possession of a head-wheel and in the kind of tines which are used.

It resembles the adjacent models in that it is capable of adjustment to different widths of drill. The possession of a head-wheel would make it easier for the operator to check a tendency toward side to side movement and to control the depth of working. In this hoe there are three " A "-shaped tines and two half-" A's." When the implement is set for a narrow drill the points of the half-" A's " lie outside the track of the heel of the tine and as, when the tines are in work, they are not visible to the operator, there is a risk of steering too near to the row of plants and to cut up the young crop ; consequently this type of tine has given place to the " L "-shaped tine. The " A " tines are a distinct improvement on the harrow teeth ; while they break up a crust quite successfully, the smaller number of tines and their tendency to cut rather than to stir makes them less liable to clog. Inv. 1928–405.

87. GORRIE'S HORSE-HOE. (Scale 1 : 4.) Lent by The Royal Scottish Museum.

This type of horse-hoe was introduced by Gorrie, of Perthshire, in 1840.

The transverse blade which follows the head-wheel provides a good means of killing annual weeds, but it has the disadvantage of being suitable only to a limited width of drill. Of the four stirrers which follow, the two leaders have their points carried well forward while the two followers are made to a rather straighter pattern. The forward tilt of the points is apparently intended to make them draw into the soil and so hold the hoe down to its work, and the flatter curve of the hinder tines is to prevent any tendency for the hoe to draw too deeply behind, and so to throw the head-wheel off the ground and to bring the cutter to the surface. The tines would break up the top layer, shake the weeds free from soil and bring them to the surface.

Inv. 1928–348.

88. LEVER HORSE-HOE. (Scale 1 : 4.) Lent by Messrs. R. Garrett and Sons.

This model was shown at the 1851 Exhibition, and embodies inventions patented in 1842 and 1849. It contains fourteen hoes each carried on a separate lever so arranged that any hoe will pass undamaged over any large stones ; two hoes pass

59

through the space between each row. To prevent damaging the crop, each hoe and its lever are adjustable so that they may be accurately pitched to suit the sowing, and the two travelling wheels for the same reason are carried on separate axle-trees so that the gauge can be adjusted. The hoes are also irregularly arranged so as to minimize the chance of tearing up the crop, and, by a lever that rotates two eccentrics carried on a shaft, the whole of the hoes can be instantly elevated clear of the ground when desired. To correct any irregularity in the movement of the horse pulling the machine, the hoe-frame is carried on swinging links, and by a shaft brought to the back of the machine the attendant can move the set sideways in either direction, so as to keep the holes accurately between the rows of young plants. Inv. 1894–161.

89. EXPANDING HORSE-HOE. (Scale 1 : 4.) Lent by The Royal Scottish Museum.

This type of horse-hoe was introduced in the middle of the nineteenth century. It is used for stirring the soil and killing weeds between the rows in a potato crop or in root crops, and for this purpose is armed with eight harrow teeth or stirrers and two " L "-shaped tines. The six leading teeth would loosen the soil and break any crust which may have formed, the tines would cut off or drag out the weeds, and the two following tines would clear the soil from the weeds and leave them on the surface to be killed by the sun. It is possible that the leading harrow teeth, on a weedy soil that was heavy and not too dry, would clog too easily and in breaking up a crust there might be a tendency to push the crust aside and endanger the young plants in the rows. The implement is adjustable so that it can be used between rows which are set near together or far apart. Inv. 1928–406.

90. EXPANDING HORSE-HOE. (Scale 1 : 4.) Lent by The Royal Scottish Museum.

This implement would be used for cultivating between the rows of the root crop after the plants have become established. It was introduced in the middle of the nineteenth century, and is adjustable to a wide range of widths of drill.

The five " A " tines would cut weeds below the surface and clean the soil quite effectively, but the mounting of tines of this pattern on the outside might make the use of the implement a little dangerous with a crop in the seedling stage. The risk of damaging plants with the outside edges of the tines would be practically eliminated where, for example, the individual plants of a turnip crop had leaves eight or nine inches long. In fact in such a case tines such as these, which can stir underneath the overhanging leaves without bruising them, have a definite advantage. Inv. 1928–404.

91. WILKIE'S EXPANDING HORSE-HOE. (Scale 1 : 4.) Lent by The Royal Scottish Museum.

This type of horse-hoe was introduced into Scotland a little before the year 1857. It is built so that expansion of the hoe to fit any width of drill can be carried out easily and rapidly, and it carries a leading " A "-shaped tine, together with a half-" A "-shaped tine at either side, one somewhat in advance of the other. These cutting tines are followed by harrow teeth for stirring. It will be noticed that when the implement is expanded for a wide drill the points of the half-" A " tines lie outside the lines of the stalks of the tines. This makes the use of the implement somewhat risky between rows of seedling plants, but becomes an advantage when the leaves have grown sufficiently to overhang the space between the rows. Inv. 1928–347.

92. WILKIE'S IMPROVED EXPANDING HORSE-HOE. (Scale 1 : 4.) Lent by The Royal Scottish Museum.

This type of horse-hoe was introduced by Wilkie in 1865.

It is evident that he has attempted to correct the outstanding faults of his earlier hoe. He has deleted the following harrow-teeth finding no doubt that the three leaders are sufficient. He has improved the shape of his leading " A " tine and has provided an ingenious expanding mechanism which permits the outside edges of the half-" A's " to lie in the line of travel when the hoe is adjusted for any width of drill. Inv. 1928–403.

93. EXPANDING HORSE-HOE. (Scale 1 : 8.) Lent by Messrs. Barnard and Lake.

In this implement, patented by Mr. F. C. Lake, in 1877, there are five shares so mounted that the total width of ground acted upon can be readily altered. The front share is fixed to the frame, while the four others are secured to two arms pivoted near the front and capable of being opened outwards by means of a pinion gearing with two curved racks attached to the stems of the hindermost shares. The stems are mounted in sleeves, and, by means of the curved arms at the top, keep the shares pointing straight whatever be the extent to which the machine is opened. When set to the desired width the parts are clamped rigidly together by tightening a few nuts.

Inv. 1891–155.

94. LINK LEVER HORSE-HOE. (Scale 1 : 8.) Lent by Messrs. Barnard and Lake.

This model represents a machine, patented by Mr. F. C. Lake in 1887, for hoeing the spaces between the rows of a crop that has been sown by a drill. It consists of a frame carried on two wheels, to which is attached behind, a sliding frame carrying ten hoes. This sliding frame is maintained accurately in its true position relatively to the crop by means of a steering rod, attached to the main frame and passing through an eye on the hoe frame. The depth of the hoes is regulated by a worm gear that, by means of two levers and chains, lifts the whole of the sliding frame, and a lever stopping into a quadrant gives an adjustment for their inclination. The hoes can be lifted clear of the ground by the wooden handle at the back, and a catch is provided that will retain them in this position. Inv. 1891–156.

HARVESTING MACHINERY

95. SICKLE. Presented by Dr. H. S. Harrison.

This example, which came from Sound Weisdale, Shetland Islands, has a very narrow blade, and the serrations are much more pronounced than in No. 96.

Inv. 1930–101.

96. SICKLE. Presented by C. Garrad, Esq.

The example shown is of early nineteenth century date and came from Romford, Essex ; it has a narrow blade, the upper surface of which has serrations inclined towards the handle giving the edge a notched outline.

The modern sickle has a wider blade than its prototype, whilst it has a plain knife edge. Inv. 1911–131.

97. FAGGING STICK. Presented by C. A. Evans, Esq., A.R.C.A.

This kind of implement has been used in conjunction with the sickle in some parts of the country for centuries.

It may be used in either of two ways. The man who is cutting the corn with the sickle may carry the fagging stick in his left hand and use it to gather the corn into sheaves as it is cut. A fagging stick is often carried by the man who is tying up sheaves after a sickle, a scythe or a reaper. He carries the stick in his right hand and gathers the corn into sheaves with it, by pulling the corn towards his left hand, or, when many thistles are present, the corn is collected against the operator's legs.

Inv. 1928–1091.

98. SUGAR CANE BILL. Lent by The Royal Scottish Museum.

This is a nineteenth-century East Indian hand implement comparable to the sickle.

It was used for cutting sugar cane, and is said to have been made from a sacrificial knife. Inv. 1928–447.

99. BELL'S REAPING MACHINE. Made by the Rev. Patrick Bell, LL.D.

This is the original reaping machine invented in 1826 by Dr. Bell, but before it was removed to South Kensington in 1868 the cutting mechanism had been altered to the later form now on the machine. The original construction is shown in an adjacent model, in which the cutting is performed by a series of double-edged shears, with the moving blade in each pair actuated by a common rod driven by a crank worked from the travelling wheels. Portions of these shears are shown on the pedestal of the reaper. The machine was a success and is recorded to have cut on an average 12 acres per day.

The machine has an open timber framework, carried on two large driving wheels with wide tyres, and two small wheels in front that directly support the cutting blades and so secure uniform clearance. It was propelled by two horses walking behind it and exerting their effort upon a pole that pushed the machine before them ; steering was done by the driver, who held a bar connected with the horse-pole.

In front is a revolving frame or collector, which supported the corn, and when cut forced it backwards on to the inclined travelling apron by which the corn was delivered at the side of the machine. The apron driving gear is reversible so that the machine could work either-handed. Inv. 1868–15.

100. REAPING MACHINE. (Scale 1 : 48.) Presented by Rev. Patrick Bell, LL.D.

This is a model of the original reaping machine, made in 1826, by the Rev. Patrick Bell.

The construction can be more clearly seen in the machine itself, but in this model the original arrangement of shears is preserved ; that on the actual machine is of much later date and was fitted by the inventor's brother. Inv. 1868–14.

101. McCORMICK'S ORIGINAL REAPER. (Scale 1 : 6.) Presented by The McCormick Harvesting Machine Co.

The machine represented appears to have been constructed and worked in 1831 in Virginia, U.S.A., by Cyrus H. McCormick, who, in 1834, patented, amongst other details of the reaper, the use of a vibrating blade, operated by a crank and having either a smooth or a toothed edge, working between fixed wires, above and below, to support the straw while being cut.

The machine consisted of a timber frame, drawn by a horse on which the driver rode ; it was supported on two wheels, the larger of which carried most of the weight, and, by bevel gearing, drove a crank that reciprocated a horizontal blade provided with cutting teeth ; the blade worked in slots through a number of spearheads, or fingers, attached to the front of a low platform, so that the fingers supported the stalks of corn while the reciprocating blade cut them through. To prevent the stalks from being pushed forward as the machine advanced, a revolving four-armed frame, or reel, driven by leather belting, was provided, by which the standing corn was supported and that which was cut was compelled to fall back upon the platform, from which it was raked in quantities suitable for forming separate sheaves by an attendant walking at the side. A protecting loop was added to deflect the standing corn from the side of the horse, and on the other side of the machine was a divider which separated the corn to be cut from that beyond ; a canvas dividing screen was also provided to prevent the cut corn from falling off the wrong side of the platform.   Inv. 1901–1.

102. McCORMICK'S REAPER OF 1847. (Scale 1 : 6.) Presented by The
      McCormick Harvesting Machine Co.

This shows McCormick's machine as improved by his patents of 1845–47 ; the width of the cut had also been increased from 4 to 6 ft. and provision made for the use of two horses in haulage.

The cutting blade was supported at the back and below almost to the edge, and the spearheads or fingers did not reach below it. By the use of an additional lever the driving gearing was brought forward so as to leave room for a seat for the attendant, who cleared the platform, while the driver rode on one of the horses. The dividing arrangements on each side were also further developed, and means were added for altering the height of the reel to suit the length of stalk being cut.   Inv. 1901–2.

103. McCORMICK'S REAPER OF 1851. (Scale 1 : 6.) Presented by The
      McCormick Harvesting Machine Co.

This represents the machine shown at the 1851 Exhibition, where it was one of two which created great interest and led to the general introduction and use of reaping machines in this country.

It differed from the earlier examples in that both the driver and raker were seated on the machine, and also in having the reel driven by pitch chain. The alterations necessary for suiting corn of different heights were provided for by so constructing the reel that its diameter could be quickly changed. Provision was also made for altering the height of the cutting edge from the ground, both the main wheel and the outer wheel being carried on adjustable swinging arms and suitable modifications made in the gearing.   Inv. 1901–3.

104. DRAY'S REAPER. (Scale 1 : 4.) Presented by J. M. Taylor, Esq.

This model is a very good representation of Dray's reaper, which, with Bell's and McCormick's, became famous during the years following the 1851 Exhibition.

The machine represented was drawn by two horses, which were driven by a man walking on their right-hand side. A second man sat on the timber housing which covered the driving wheel. This man placed his right foot in a loop on the foremost cross-piece of the frame, and his left foot in a loop attached to the tilting board on which the cut corn collects. With the aid of a rake he assisted the fall of the cut grain on to the board, and by placing the rake behind the pivot of the board, pressing with the rake and lifting with his left foot he was able to deliver the cut corn in sheaves.

The model is interesting in its similarities to the modern machine. The cutting apparatus of Dray's or Hussey's machine is not very far removed from the modern idea. The lay-out of the mechanism which conveys motion from the driving wheel to the knife is, in this model, not greatly different to that of Hussey's machine, and its close relationship to that of the modern mower is striking.   Inv. 1929–888.

105. SELF-RAKING REAPER. (Scale 1 : 8.) Presented by Messrs. Harrison
      McGregor and Co., Ltd.

The model shown represents a type of reaper made in the decade 1860–70, and employed until recent times.

The cutting mechanism is of the ordinary type, but the driving mechanism and the attachment of the driving wheel to the main frame is somewhat primitive.

The corn is cut close to the ground by the knife and falls backward on to the plat-form, where it collects until one of the arms of the reel sweeps the platform clear, so depositing the corn on the ground in sheaves. The path of the arms is so arranged that when they rotate forward from their vertical position, they brush the corn on to the platform until they have just passed the knife. From that position they may either descend and sweep the platform or if the driver depresses the lever at his foot the arms will return to the vertical without dropping on to the platform. In this way the driver has some control over his sheafing apparatus ; he is also able to throw the mechanism in or out of gear by using the lever that lies forward of his seat and on the other side of the wheel. The height at which the corn will be cut is controlled by two adjustments. First, the general height of the machine from the ground is set by using the clip on the right-hand wheel and turning the horizontal hand-wheel by the driving wheel ; this would be done while the reaper is stationary. Adjustment of height during working is effected by using the lever that is set somewhat far back by the driving wheel.   Inv. 1926–932.

106. JONES'S MOWER. (Scale 1 : 8.)

This is a model of an American mower which was popular in the British Isles at the beginning of the twentieth century. It differs very little in general principles from the modern mower. The drive is taken through a dog-tooth clutch on the main axle to a large bevel gear-wheel, which drives a smaller one mounted on a short shaft. The shaft lies at right angles to the axle and carries a pinion at its forward end on which is carried an eccentrically mounted boss. A connecting rod joins the rotating boss to the knife head and imparts to the latter reciprocal motion. The blades of the knife working from side to side in the fingers of the knife-bar constitute the cutting mechanism.

The driver sitting in the seat can reach and control with his left foot a small lever which, by moving the clutch, can throw the mechanism in or out of gear. The upright lever at his right hand on being pulled backwards will cause the knife bar to pivot on its left hand end and assume a vertical position. This control is necessary in order to avoid smashing the knives and fingers against stones, ant-heaps or mole-heaps. There was also a third control, a lever which lay to the driver's right hand, by means of which the knife bar could be rotated, throwing the points nearer the ground or raising them, thus regulating the height at which the grass was cut.   Inv. 1928–589.

107. BUDDING'S ORIGINAL LAWN MOWER. Lent by Messrs. Ransomes, Sims and Jefferies, Ltd.

This type of lawn mower was patented by Edwin Budding in 1830, and was described as a new application of machinery for the purpose of cropping or shearing lawns, etc. It is evident that it was the first attempt to replace the scythe for this purpose.

The manufacture of this machine was taken up in 1832 by Messrs. J. R. and A. Ransome, the firm now known as Messrs. Ransomes, Sims and Jefferies, Ltd.

The machine shown, to be complete should carry a flat box low down in front of the cutting apparatus to catch the cut grass. It differs very little from the description in the patent specification except that the original clutch is slightly different and the material of the small-diameter roller was specified as cast iron. The roller is mounted on a shaft which drives, through the clutch, a large gear-wheel. This turns a sprocket on the intermediate shaft, which gives motion to the gear-wheel to the left-hand side. This gear-wheel meshes with the sprocket which drives the knives. The knives rotate at about twelve times the speed of the roller and obtain their cutting action by working against the rigid knife bar on the underside of the machine.

Budding concludes his specification with the following :—" Grass . . . too weak to stand against a scythe . . . may be cut by my machine as closely as required, and the eye will never be offended by those circular scars, inequalities and bare places so commonly made by the best mowers with the scythe. . . . Country gentlemen may find in using my machine themselves, an amusing, useful and healthy exercise."
                                                                                          Inv. 1926–808

108. TWO STACK LADDERS. (Scale 1 : 8.) Lent by The Royal Scottish Museum.

These two models illustrate the kind of ladders that are constantly in use on the farm. Both are needed during the making of a hay-rick, or a corn-rick, and the longer ladder is part of the necessary equipment required for thatching ricks.
                                                                                Inv. 1928–425, 429.

109. STRAW-ROPE SPINNER. (Scale 1 : 4.) Lent by The Royal Scottish Museum.

This is a model of a mid-nineteenth century instrument for spinning straw ropes, four at a time.

The use of straw ropes for tying down thatch was once universal, but toward the end of the nineteenth century straw was being replaced by yarn, although up to the present day the older practice persists in outlying districts. In Scotland the making of a straw rope required two people. One of them carried a crook or an implement like a carpenter's brace and bit, with a hook instead of the bit. The other person sat on a heap of straw, placed a wisp of straw over the hook, and while the assistant walked slowly backward twisting the brace, the other let out the straw gradually from his left hand and supplied small portions of straw in equal quantities with his right. The operation requires considerable care and skill, if the rope is let out unequally it breaks too easily at the narrow parts, or at the places where it is twisted too tightly. If it is not twisted sufficiently it comes apart when it is pulled, and if the twister does not walk backward sufficiently quickly to keep the rope taut it runs up into exceedingly obstinate loops. The model is designed to allow four rope-makers to work with one twister. Inv. 1928–458.

110. RICK BORER. (Scale 1 : 5.) Lent by Messrs. Workman and Sons.

This is a model of an instrument used for cutting a circular hole downwards into a heated haystack, that will allow the external air to reduce the temperature by ventilation.

It consists of a shell cutter with a long square shank, that can be further lengthened by extension rods. A cross handle that can be slid along the rods is provided for rotating the cutter by hand. Inv. 1894–159.

111. HAY KNIFE. Lent by The Royal Scottish Museum.

This implement is used for cutting hay out of the rick.

When the rick has stood for some weeks the hay settles down very tightly and when it is needed for use, to be fed to stock, an instrument with a sharp heavy blade is necessary to cut it. Inv. 1928–430.

112. HAYMAKING MACHINE. (Scale 1 : 8.) Lent by Messrs. Robert Boby, Ltd.

This model of a hay-maker, or " Tedder," represents one of the earlier machines which were introduced into the process of hay-making, and were in use in the early and mid-nineteenth century.

The machine consists of a pair of spinners mounted on a transverse spindle, driven from the ground wheels and revolving in the same direction as the wheels and at high speed. Inv. 1914–23.

113. HAY TEDDER. (Scale 1 : 8.) Lent by The Massey-Harris Co.

This model shows a modern form of machine for tossing hay. The hay is thrown over by six forks, which, instead of the rotating motion adopted in the other types of hay-making machine, have a kicking motion closely resembling that of a hay-fork used by hand.

The machine has a timber frame terminating in shafts so that it can be drawn by a horse, the forks working behind. Two large carrying wheels, running loose on their axles, support the machine, and each carries a spur wheel which by an intermediate wheel drives a three-throw crank to the pins of which the middle of the fork handles is attached. The upper end of the handle is held by a link connecting it with the framing, each of the driving wheels working three of the forks. A seat is provided for the driver from which he can control the distance of the forks from the ground and can also throw the tedding motion in and out of gear. Inv. 1893–184.

114. SWATH TURNER. (Scale 1 : 4.) Made by Martin's Cultivator Co.

This represents a machine used for turning over the swaths of hay while it is being cured. It was patented by Mr. W. E. Martin in 1903.

The machine is carried on two wheels which drive two revolving rakes set obliquely to the direction of motion and having spring tines which always remain vertical ; the mechanism employed is an application of Boehm's method of coupling two parallel shafts by means of a number of inclined links or coupling rods. The wheels run loose upon a transverse axle fixed to the framing and from this axle two bars project straight forward ; they are then bent horizontally through 60 degrees, and then forward for a short length where their ends are supported in bearings whose height is adjustable. On each of these bars, close up to the ends of the oblique portion, two hubs are mounted having four arms, the corresponding ends of which are coupled by four oblique rods, whose parallel ends bear in holes in them. The tines are fastened to these rods and always hang downward. The rear hubs are driven from the carrying wheels by a sleeve and bevel gearing, through clutches operated independently by hand-levers. The heights of the turners are adjusted by hand-levers, while foot-levers are also provided for suddenly raising them when passing an obstruction. Stripping bars are fitted to prevent the hay from being carried round ; solid discs are fitted behind the turners and a baffle plate between them. A seat is provided for the driver.

Inv. 1907–18.

115. WOODEN DRAG-RAKE. Presented by Arthur H. Ogilvie, Esq.

This is a type of rake which was used from very early days before the introduction of horse-rakes in the nineteenth century, and similar rakes still have their uses to-day.

It is made throughout of wood and is used for raking up after the bulk of the hay crop or the sheaves of the corn crop have been cleared. The worker grasps the end of the handle with one hand and the crosspiece with the other and walks along with his back to the rake, lifting it when it is filled.

Inv. 1928–1201.

116. IRON-TOOTHED DRAG-RAKE. Presented by Arthur H. Ogilvie, Esq.

This type of rake is of quite early origin, but is used fairly widely at the present day for work which does not require a horse-rake.

The handle, which is a replica, is of wood and shaped somewhat like an " A." It is used nowadays chiefly for raking the corn stubble after the sheaves have been cleared.

Inv. 1928–1202.

117. AMERICAN HORSE-RAKE. (Scale 1 : 8.) Lent by The Royal Scottish Museum.

This type of rake, imported from America in 1830, was used for running hay up from the swath into windrows, for gathering windrows into heaps or " cocks," and for collecting hay and dragging it to the rick when the hay was stacked in the same field.

The rake is drawn by a single horse and the tines are locked in their present position so that hay is collected by the set of tines which are pointing forward and downward. The staves, which are attached to the handles and which run down vertically, are joined at their lower end by a roller. This roller lies on the idle set of teeth and prevents them from rising. When the collecting teeth are full, the handles are raised ; this permits the forward points to hold on the ground and the teeth can then rotate round the main bar, so emptying a rake-full of hay and presenting the other set of teeth for collecting the next load.

Inv. 1928–415.

118. SINGLE-HORSE HAY SWEEP. Presented by Arthur H. Ogilvie, Esq.

This type of sweep was introduced from America in 1850. In the original design the tines were made of wood, iron being introduced at a later date.

The horse pulls the sweep by a fairly long chain which is connected to the rings at the ends of the two side chains. The driver walks behind holding the reins and the two wooden handles. The tines lie at a low angle to the horizontal, and hay is collected on the set of tines which point forward. The rake is prevented from rotating by the fact that two of the backward pointing tines are held down by a pair of small wooden bosses on the inside of each handle. When the forward pointing set has collected a load of hay, the sweep and its load may be driven a short distance across the field to the rick or by separating the handles slightly the backward-pointing tines are released and the rake rotates, emptying itself and presenting a fresh set of tines to be filled.

Inv. 1928–1109.

119. HAY ELEVATOR. (Scale 1 : 8.) Lent by Messrs. G. H. Innes and Co.

This appliance for lifting hay, straw, etc., and depositing it in a high heap so as to form a rick is a development of a machine first introduced about 1850. It consists of a wagon on which is mounted a long inclined trough in which work two endless pitch chains connected by transverse bars fitted with prongs or forks which carry the material from an adjustable hopper at the bottom of the trough and deliver it over the centre of the site of the stack ; the chains are driven by sprocket wheels from a horse gear or by belting from a portable engine. The upper end of the trough is supported by means of a bar which slides under while it is being elevated by two hinged props, combined with annular wheels worked by pinions and winch handles. The hopper end of the trough is mounted on a hinged frame which can be gradually raised, and at the same time moved outwards by means of a rope and winch handle. As the stack increases in height this enables the delivery to be always near the centre and the stack can be topped up without any pitch hole. The whole machine can be folded up into a small space for travelling.

In a modification of the machine, the hopper can be lowered until the forks touch the ground when the hay, etc., can be pitched directly from a sweep rake on to the stack.

Inv. 1904–116.

66

120. 15-TON ENSILAGE STACK PRESS. (Scale 1 : 12.) Lent by The
Aylesbury Dairy Co., Ltd.

When it is impracticable to dry hay in the open the crops may be stored wet,
without being liable to spontaneous combustion, if sufficiently compressed. This
may be done in a silo or in a stack press as shown in the model. The rick is built
over timber logs laid in the ground and projecting beyond its sides. Before thatching
galvanized wire ropes are carried over the top of the stack and down to small winches
secured to the timbers. These winches by means of a long lever enable a heavy
pull to be exerted upon the ropes and also form a ready means for continuing the
pressure as the ensilage becomes more compact. Inv. 1890–89.

121. INDIAN RAKES. (Scale 1 : 4.) Lent by The Royal Scottish Museum.

These are models of nineteenth century rakes used in India.

Their primitive form may be considered to indicate the kind of implement which
was used in this country before the commencement of the use of metal teeth, and the
subsequent introduction of horse-rakes in the first quarter of the nineteenth century,
Inv. 1928–450, 527.

THRESHING MACHINERY

122. THRESHING IMPLEMENTS.

This collection, consisting of a flail, a replica of a piler, and a fan, represents the implements used in threshing and cleaning grain before the advent of the threshing machine. Implements of this type were in use until toward the end of the nineteenth century, particularly in remote country districts.

The sheaves of corn were laid on the floor of the barn in two rows with the ears of the corn inward and the butts of the sheaves outward. The flail was then used to beat the heads of the sheaves in order to separate the grain from the straw. When barley was threshed with the flail it was not possible by this means alone to break off the awns. The threshed barley was laid on the floor of the barn and was treated by working the piler up and down in the heap of grain.

The grain at this stage would have mixed with it a large amount of chaff and bits of broken straw. To remove these, the mixture of chaff and grain was spread on the floor in front of the fan, and while the fan was turned by one man, another man turned the heap with a shovel, the draught created by the fan blowing the lighter chaff away and leaving the heavier grain.   Inv. 1909–127, 1928–290, 1921–526.

123. NORAG.

This implement is still used in Egypt for threshing straw crops. An implement somewhat similar to this one was used in Italy and Spain, and it is thought that the idea was introduced into North Africa by the Romans.

The crop, after it has been harvested, is spread fairly evenly and thickly on the ground, and the norag, drawn by oxen, is driven about over it. The discs on the rollers, cut the straw into short lengths ; and the rubbing of the sledge and the treading of the oxen free the grain from the ear. After this treatment there remains a heap which comprises a mixture of grain, chaff and short straw. The straw is removed by men armed with forks. The grain and chaff are separated by a winnowing process which makes use of the wind. The mixture is scooped into a basket, which is lifted above the operator's head and tilted so that the contents fall to the ground. The heavy grain falls at the operator's feet, while the light chaff is carried away on the breeze.   Inv. 1924–323.

124. PHOTOGRAPH OF NORAG. Presented by Col. Sir Henry Lyons, D.Sc., F.R.S.

This photograph shows the Norag in use. The implement drawn by a team of oxen is being driven about over the heap, while men with forks are removing the straw. The winnowing or separation of chaff and grain would be carried out later.
                Inv. 1915–576.

125. THRESHING BOARD. Presented by the Government of Cyprus.

This type of board is used in Cyprus at the present day for threshing grain.

It is closely allied to the Norag used in Egypt, and is the kind of implement which was used generally in Mediterranean countries up to a century ago. The harvested crop is spread evenly on the ground and the board, drawn by a bullock team, is driven about over it, thus bringing about by rubbing, the separation of the grain and chaff from the straw. The extraction of the straw and the process of winnowing the chaff from the grain are carried out later as separate processes.   Inv. 1929–636.

126. TOOLS. Presented by the Government of Cyprus.

These tools are used for inserting flints into the threshing board.

The larger hammer is used for breaking the flints, the smaller one for hammering the flints into the board. The mallet and chisels are used for making the incisions in the board, the holes are cleared with the fourth chisel, and broken pieces of flint are prised out with the spike.   Inv. 1929–894.

127. PILER. (Scale 1 : 8.) Lent by The Royal Scottish Museum.

This is a model of the piler shown with the primitive threshing implements.
                Inv. 1928–503.

128. BARLEY HUMMELER. (Scale 1 : 4.) Lent by The Royal Scottish Museum.

This machine acts as a mechanical piler for breaking off the awns from grains of barley.

The belt pulley at the top of the machine drives a shaft whose motion is communicated through bevel gearing to the vertical shaft. The latter passes through the bottom of the receiving hopper and extends almost to the bottom of the cylinder. At its lower end blades are attached, so that when grain is run into the cylinder from the hopper, the rotating blades stir the grain with a certain amount of violence thus breaking off the awns. If the hopper is kept filled the extent to which the grain is treated can be adjusted by altering the speed with which the grain passes through the trap at the bottom of the cylinder. Inv. 1928–459.

129. BARLEY HUMMELER.

This is an example of a piler made in the form of a roller.

The implement is used for breaking the awns from barley. The grain is laid out in a heap on the floor of the barn and the piler, or hummeler, is drawn to and fro over it. Inv. 1928–1085.

130. STRAW SHAKER. (Scale 1 : 4.) Lent by The Royal Scottish Museum.

This model represents that part of a threshing machine which is concerned with shaking the grain free from the straw. It was submitted to the Highland Society and was awarded a premium in 1837.

When the machine was at work the threshed straw, grain and chaff, after passing through the drum, was flung on to the shakers. The straw worked its way along slowly and fell out of the machine at the far end, while the grain and chaff would drop between the shakers and lie underneath the machine ready to be shovelled up and put through the winnower. Inv. 1928–457

131. PORTABLE HORSE-POWER THRESHING MACHINE. (Scale 1 : 4.) Lent by Messrs. R. Garrett and Sons.

This model was shown at the 1851 Exhibition, and embodies inventions patented in 1843–1850. The drum has five straight iron blades as beaters, and the " concave " is of ribbed iron plates separated by spaces covered with wire screens. The corn is threshed out of the straw by the rapidly revolving beaters, which knock out the grain as the straw is being carried round in the small space between the drum and the concave, the empty straw finally being ejected at the lower end of the drum. Here the straw passes over a wooden grid which recovers any short ears that may have been carried round unthrashed. The grain threshed out drops through the concave into the space at the bottom of the machine, and is removed at intervals through the side doors provided. The concave is hinged and the clearance is readily adjustable by screws.

The corn from these machines was afterwards finished in a corn dresser, but self-acting straw shakers that removed any grain remaining in the straw were frequently fitted to such threshers. The machine was driven by animal power, steam or other means ; the horse gear is arranged for four horses which, working at the rate of two miles per hour or three rounds per minute, are stated to have threshed as much as 60 bushels per hour. To render the machine portable, the horse gear is fitted with an axle having two road wheels ; the threshing machine can be placed upon the horse gear, so that the whole apparatus travels on two wheels. Inv. 1894–160.

132. WINNOWER. (Scale 1 : 4.) Lent by The Royal Scottish Museum.

This is a model of a type of machine, introduced in the eighteenth century and used for separating grain from chaff.

This winnower probably resembles the older Dutch machine, as, by 1830, winnowers had been designed to remove weed seeds as well as chaff, an improvement which was effected quite early by one of the Meikle family. In this model, there is a fan which is turned by hand, and the blast of air which it creates is directed across the stream of mixed grain and chaff as it falls vertically from the hopper. The lighter chaff is blown clear of the machine and the grain falls and is shot out to the side. The rate of flow of the grain is regulated by a swinging door at the bottom of the hopper and the force of the blast would be adjusted by the rate at which the handle was turned. Inv. 1928–359.

133. WINNOWER. (Scale 1 : 4.) Lent by The University of Edinburgh.

This is a model of an improved winnower of the early nineteenth century which in addition to removing chaff also separated out weed seeds.

The arrangement of the hopper and the fan is similar to that of the older pattern, but in this case the stream of grain and chaff is directed on to a pair of rocking screens which hold up everything larger than the grain. The blast is directed upward through the screens so that as before the chaff is blown clear of the machine, but thistle heads and other large weed seeds held by the sieves are shaken forward into a shute which leads them to the side of the winnower. The grain falling through the sieves is caught on a broad inclined plane into which is fitted a stationary sieve whose mesh is sufficiently small to pass small weed seeds and to hold the grain. This slanting sieve allows the grain to roll down and emerge from the machine under the fan casing.

The mechanism is not complicated. The shaft of the fan carries a small crank which is joined to a lever by a connecting rod running horizontally along the side of the machine. The lever is attached to the rocking sieves which are pivoted at the end nearest the fan, thus when the handle is turned the fan is rotated and the sieves are swung from side to side. *Inv.* 1928–638.

134. WINNOWER. (Scale 1 : 6.) Presented by J. M. Taylor, Esq.

This is a model of a late nineteenth century winnower. It differs from earlier models in having a more efficient arrangement of screens and a more powerful blast.

The machine is worked by hand ; the fan is driven, through an external spur gear, from the handle shaft. Connecting rods run from the pinions on the fan shaft to the pair of upper screens and give them a backward and forward motion. This reciprocal motion is communicated to the lower screen by means of a pair of levers which are pivoted at their centres, and whose upper ends are attached to the upper screens.

A mixture of grain, chaff, weed seeds and short bits of straw is tipped into the hopper at the top of the machine. This mixture first meets the upper screens, and their interstices are just large enough to allow the grain to pass through them. The blast of air which plays across these screens carries the chaff straight out of the machine. The bits of straw and the larger weed seeds are held by the screens ; they work their way along the screens assisted by the blast, and finally fall down the shoots to the sides. Grain and small weed seeds fall on to the lower screen. Here the weed seeds pass through the screen, collecting under the machine and the grain is held by the screen and passes out below the fan casing. *Inv.* 1929–1048.

135. BEATER BARS AND DRUM ENDS FOR THRESHING MACHINE. Lent by J. Goucher, Esq.

These important details are usually made of malleable cast iron, so as to obtain sufficient toughness without entailing any great cost in manufacture. The beater bars are fixed longitudinally on wooden bars that connect the drum ends shown, and the complete drum revolves at about 1,200 revs. per min., close to the " concave " or grid between which and the drum the straw to be threshed is fed.

The form of bar shown was patented in 1848 by Mr. Goucher and has now become widely used. On its face are diagonal channels which reduce the breakage of the straw and grain, probably owing to the certainty with which the straw is carried round and also to the clearance spaces provided for the grain. One of the bars has been bent to show the toughness of the metal employed. *Inv.* 1862–29.

136. THRESHING MACHINE. (Scale 1 : 4.)

This sectioned model represents the form of double-blast machine made in 1860 by Messrs. Wallis, Haslam and Steevens. The platform and other boards have been placed in the position for working, and not as they are when the machine is closed for travelling.

When threshing is in progress, the sheaves of corn are opened and placed on the platform, while an attendant, standing in the closed recess, or " dickey," in front of the hopper, or " mouth," throws the corn, at an angle of about 45 degrees, into the space at the bottom of the hopper ; here it is caught by the ribbed " beaters " of a drum making about 1,000 revs. per min., which carry it round through the space between them and a curved grid, or " concave," which delays its passage to such an extent that the grain is knocked out of the ears of corn and passes through the grid, while the long straw is carried round and delivered at the opposite edge. The grain, chaff, and short straw, after passing through the bars of the concave, slip down an inclined board and are delivered at the higher end of a sloping sieve, or " caving riddle," which is rapidly reciprocated longitudinally, so that the corn and chaff fall through the holes in the riddle, while the larger pieces of broken straw pass along and are delivered at the end, where they fall on to the " caving board " for removal. The corn falls on to a lower vibrating frame carrying a board on which it travels till it drops over the edge on to a finer screen reciprocating with it ; through the meshes of this " chaff riddle " a blast of air, from a revolving fan, is continually rising, so that any

70

kind of grain, owing to its density, passes through the sieve, while chaff and other light particles are blown away.

The grain now moves nearly horizontally to the side of the machine, where by a cup elevator it is raised to the level of the top ; it is then allowed to fall in front of the blast from another fan, by which any remaining light particles are removed and the grain separated, by a dividing board, into two grades which pass into separate sacks attached to external hooks.

The long straw thrown out by the drum falls on to three inclined grids or " shakers," which are reciprocated by a three-throw crankshaft, so arranged that the shakers, while tossing the straw and thus releasing any entangled grain, slowly work it out at the end of the machine, where it falls on to the " straw board " ready to be carried away for farm use ; any grain passing through the shakers is received on a vibrating inclined board, down which it slides to the other grain coming from the concave.   Inv. 1900–32.

137. SHEARER'S THRESHER. (Scale 1 : 4). Lent by The Royal Scottish Museum.

This is a mid-nineteenth century thresher. It differs from Ransome's somewhat in design and was intended to be built in the barn. It is rather narrower at the drum and requires about 6 h.p. to drive it whereas Ransome's thresher being capable of a higher output would probably need about 10 h.p.

The bars of the beaters on the drum are armed with short sturdy spikes. The concave is closed by a sheet of iron so that as well as the straw, the grain and chaff are thrown on to the shakers, which are of the normal modern pattern. There is no upper shoe, and the grain, chaff, and short lengths of straw fall out of the shakers, straight on to the caving riddle. Here the straw is stopped and grain and chaff fall through and are caught by the lower shoe. The caving riddle and lower shoe are both rocked forward and back, throwing the cavings forward and dropping the grain and chaff into the first dressing shoe. Here the chaff is blown clear, and a single sieve separates the large weed seeds and allows the mixture of grain, broken and shrivelled grains and small weed seeds to fall into the elevator trough. The mixture is then carried up at the side of the machine and falls into the second dressing shoe, where the processes of the first dressing are repeated. Larger weed seeds are separated and thrown clear, and the grain and small weed seeds pass out at the side of the machine. The separation of small weed seeds and small and broken grains is carried out as a separate operation with a special machine—a seed cleaner.   Inv. 1928–551.

138. RANSOME'S THRESHING MACHINE. (Scale 1 : 4.) Lent by The Royal Scottish Museum.

This is a type of thresher which was being built in the latter half of the nineteenth century. It was intended for work with a portable steam engine and was lightly built and fitted with shafts so that it could be moved by horses from one stackyard to another or from farm to farm.

The machine threshes the crop, clears the grain from the straw, and separates the chaff and weed seeds from the grain, but does not remove broken or shrunken grains from the bulk.

The sheaves of corn are fed into the machine at the top and are rubbed between the eight plain beaters of the rotating " drum " and the bars of the " concave." The concave is adjustable at three points ; at the top where it is usually set fairly well away from the drum, at the bottom where it is set close to the drum, and at the half-way point, at the join of the two halves of the concave. Should the concave be set too closely there will be a large number of broken grains, while too much room between the drum beaters and the concave will result in incompletely threshed ears passing out with the straw. Much of the grain will be thrown through the ribs of the concave and will fall with a good deal of chaff and broken straw on to the " cavings riddle." The straw is thrown on to the " shakers," which consist of 15 spiked rollers, triangular in cross-section, set parallel to each other across the machine. Grain held by the straw is shaken clear and falls between the rollers, and the straw passes along to fall clear of the machine. The " upper shoe " is really the floor of the trough in which the rollers are set, and, partly because of the angle at which the trough is hung, and partly because the shoe is close enough up to be swept by the spikes of the rollers, grain and chaff falling on it are swept back, to fall on to the cavings riddle. This riddle is jogged from side to side, and it is designed to separate broken short lengths of straw, while grain and chaff pass through it and are guided by the " lower shoe " into the " dressing shoe." The dressing shoe is also jogged, and is equipped with two sieves and a fan—it is really a winnower. The fan blows the chaff clear of the grain, the upper sieve stops large weed seeds and passes the grain, and the lower sieve stops the grain and passes small weed seeds. Large and small weed seeds fall into the tray at the bottom of the dressing shoe, and are passed out of the machine through an aperture beside the bottom of the " Elevator." The grain is shaken into the trough of the elevator, and is carried up to a point near the drum. Here it can either be shot straight into a shute which leads to the right-hand spout, or it can be made to enter a conveyor, and is taken across the machine to the left-hand spout. Both spouts are provided with hooks so that sacks can be attached to collect the grain.

Inv. 1928–547.

139. REED WHIPPER. Presented by A. H. Oglivie, Esq.

This is an example of an implement used in the nineteenth century and probably earlier, in Devonshire, for making reed for thatching.

Reed is made by threshing and combing wheat. The sheaves are taken from the stack, and combed with a short-handled long-toothed comb, when the tangled straw and some of the grain is removed. The operator then takes hold of a double handful of the combed straw, grasping it at the butt-end, and beats the heads on the bars of the reed whipper in order to remove the remainder of the grain. The reed which remains in his hands is bound up into niches or bundles weighing about 36 lbs., and the grain and chaff which collects under the whipper is winnowed. Inv. 1929–534.

CARTS AND WAGONS

140. SIAMESE SLEDGE. (Scale 1 : 10.) Presented by J. W. Hinchley, Esq.

This is a model of the carrier or sledge used in Siam for carrying the rice straw, etc., when harvested. It is drawn by a single ox or buffalo. Inv. 1914–130.

141. CORN BARROW. (Scale 1 : 4.) Lent by The Royal Scottish Museum.

This is a model of a barrow which was used in the eighteenth and nineteenth centuries for carrying sheaves from the stack to the thresher.

This kind of implement has fallen into disuse as the necessity for better organization of labour on the farm has arisen and the modern type of thresher has come into use. When the corn was threshed with a flail or a hand thresher, or in the early days of built-in, horse-driven threshers, it was feasible to wheel barrow-loads of sheaves from the rick to the barn. Nowadays, either the portable thresher is moved from rick to rick, or there is sufficient room in the barn to haul by means of horses and carts and to store enough corn for a day's work or more for the built-in thresher. Inv. 1928–426.

142. HAND-BARROWS. (Scale 1 : 4.) Lent by The Royal Scottish Museum.

These are models of nineteenth century hand-barrows which were used for the same purpose as the corn barrow (No. 141), to carry sheaves from the rick to the barn. Inv. 1928–427, 558, 559.

143. COMMON WHEELBARROW. (Scale 1 : 4.) Lent by The Royal Scottish Museum.

This model of a wheelbarrow was included in a collection of typical farm implements and machines, established by the Highland Society in 1831.

The common wheelbarrow has its uses in the farmyard and the garden for carrying food to stock and carrying dung to the midden, etc., though on some of the larger farms of the present day its place in the yard is taken by overhead bucket-ways. Inv. 1928–460.

144. SACK BARROW. (Scale 1 : 4.) Lent by The Royal Scottish Museum.

This type of barrow is used for moving sacks in the granary.

A sack of grain will weigh anything between 1·5 cwts. and 2·25 cwts., a bulk which is not easily moved about by one man without assistance. The barrow is wheeled up to the upright sack, the iron blade is pushed as far underneath the sack as possible, and the man places one foot on the axle, holding the barrow with one hand and the top of the sack with the other. He presses forward with his foot and pulls the sack toward the barrow. As the weight of the upper part of the sack falls on the barrow, the blade will be lifted off the ground bringing the sack of grain with it. Inv. 1928–428.

145. INDIAN BULLOCK CART. (Scale 1 : 4.) Lent by The University of Edinburgh.

This model cart was sent to the Highland Society from Lahore, and represents the kind of cart used in that district of India.

The construction of the cart is very simple ; two beams are set obliquely across two parallel crossbeams, the fore-ends of the oblique beams meeting and forming the draught pole. The axles of the two wheels are carried outwards from the oblique beams and the wheels are prevented from slipping off the axles by a pair of light timbers which join the ends of the crossbeams ; these light timbers also support some of the weight of the frame on the axles. Inv. 1928–632.

146. CART. (Scale 1 : 4.) Lent by The Royal Scottish Museum.

This is a model of a common farm cart, not adapted for tipping.

It is provided with a mechanism, actuated by the handle at the back, for thrusting the body of the cart forward on the frame when going up-hill and drawing the body back when going down-hill. Inv. 1928–371.

147. CART. (Scale 1 : 4.) Lent by the Royal Scottish Museum.

This is a model made in the mid-nineteenth century of a common farm cart, not adapted for tipping. Inv. 1928–370.

73

148. TIP CART. (Scale 1 : 4.) Lent by The Royal Scottish Museum.

This is an early nineteenth century model of the type of cart still used in Scotland and in the Midland Counties for general farm haulage.

It is provided with a tipping mechanism for use when carting roots, manure, etc., and a frame is available to be bolted to the sides of the cart for carrying straw, corn, or hay.   Inv. 1928–461.

149. TIP CART. (Scale 1 : 4.) Lent by The University of Edinburgh.

This is a model, made prior to 1830, of a type of general purpose farm cart, which is still popular in Scotland and the Midlands of England.

Fitted with the frame, as it is at present, it is intended for hauling hay or corn. The frame is not bolted, but simply fitted on so that it can be removed without any difficulty, and the cart is ready for hauling sacks of grain, mangolds, etc., or dung. The body of the cart is fixed to the axle and can pivot on it for tipping purposes. It is held by a pin fixed between the shafts, so that it shall not tip unless required. The shafts are attached to the body of the cart some distance ahead of the line on which the cart pivots on the axle. This is done so that the horse, by leaning back in the breeching, can assist the man in tipping, and by pulling on the shafts can assist in righting the cart after tipping.   Inv. 1928–610.

150. HAY CART. (Scale 1 : 4.) Lent by The Royal Scottish Museum.

This is a model of a common hay cart, made early in the nineteenth century in Scotland.

It is also typical of the kind of cart that is used for hauling hay, corn, etc., in many parts of the British Isles, particularly in hilly districts where the fields are small, and where the large four-wheeled wagon is awkward to handle.   Inv. 1928–416.

151. CART. (Scale 1 : 4.) Lent by The Royal Scottish Museum.

This type of cart was invented by Mr. Watt, of Biggar, and the model was presented to the Highland Society in 1827.

The cart represents an attempt to solve certain problems. Large wheels are an advantage because they reduce the draught, but large wheels and a straight axle mean awkward loading and a high centre of gravity. By using large wheels and a cranked axle, the draught is light, loading is easier, and the cart either empty or loaded is very stable. There is also the disadvantage with large wheels and a straight axle that, should the horse fall, there is a greater possibility of snapping the shafts when they hit the ground, a danger which is minimized by using the cranked axle. The two rods which run along the upper sides of the two shafts are attached to the breeching chains, so that in going down hill, when the horse checks the cart, by bearing back into the breeching, the rods are thrust backwards and the beam, attached to their hind ends, drives the brake blocks on to the wheels.   Inv. 1928–374.

152. HAY CART. (Scale 1 : 4.) Lent by The Royal Scottish Museum.

This is a model of an improved corn and hay cart, made by A. Robertson of Parkhead, Alloa, N.B., in 1835.

It is a very simple design and has the further advantage of allowing the bottom layers of the load to occupy the full breadth of the cart. By this means the centre of gravity of a load of corn or hay is lower than it would be either in the common corn cart or in the dung cart with frame, so minimizing the risk of upsetting. The absence of sides also facilitates loading.

The cart can be converted for timber work by unbolting the frame and replacing it by two cross-bars, one before and one behind the wheels.   Inv. 1928–372.

153. SUSSEX WAGON. (Scale 1 : 8.)

This is a very old type of English farm wagon, found chiefly in Surrey, Sussex, Kent and Hampshire.

The early designers of this vehicle wished to retain fairly large front wheels, without having the body too high from the ground, and the waisted front represents the best solution of the difficulty. Curved timber was specially selected from which to cut the front frame pieces. The centre beam was morticed to take the framing and also the two " summers " running throughout the length of the wagon. The coupling pole, connecting the under-carriages, continued past the rear axle and served to support the back of the wagon, besides preventing this from lifting from the under-carriage. The sides were of thin tough ash plank, nailed against ribs, which passed right through the side rails.

The wheels were generally from 4 to 6 inches in width, the tyres being often put on in separate pieces or strakes, and fixed on while hot with large oblong-headed

74

strake nails. As the wheels were dished and hung outwards at the top, the felloes and tyres were coned, to provide flat contact with the ground. In different districts village wheelwrights made wheels of varying widths to suit the local type of soil, the wide wheels being preferred for the heavy weald clays, and the narrow for hilly country, dry chalk downlands, etc. The axles were often made of wood throughout, tough beech being the favourite timber ; the arms were rounded, tapered, and shod with steel on the under side. This necessitated very large hubs to the wheels, but it was commonly held that the wooden-armed wagons with their large bearing surface within the box of the wheel made for easy running. Later the iron stub axle displaced the wooden axle, although the latter may occasionally be seen in very old wagons.

The shafts were generally fixed to the fore-carriage by a strong iron pin passing through both " hounds " and shafts. Inv. 1930–1.

154. OXFORD WAGON. (Scale 1 : 8.)

This is a model of the old English farm wagon, which may still be seen in Oxfordshire, Berkshire, Wiltshire and adjacent counties. It may be described correctly as one of the best products of the old-time village wheelwright.

Its distinguishing features are the waisted front, to allow for maximum locking of wheels in turning, the hump-backed formation of the " raves " over the high back wheels, and the upward curve of the frame in front for better clearance of the front wheels. The coupling pole underneath serves to connect the rear under-carriage with the fore under-carriage, thereby relieving the body of the wagon of the tension of the draught. The locking chains can be taken up should it be desired to restrict the extent of the locking, or traverse of the front wheels. The wheels themselves are dished, or concave, turning on axles which are bent downwards at the arms in order that the spokes may assume the vertical position as they take the load. The rear end of the coupling pole continues past the rear axle and is fitted with a cross-piece on which rests the two " summers," or inside framing of the body. A bolt or more usually a keyed round-pin passes through each end of the cross-piece, and each summer. These bolts are fitted loosely, to allow for play between the body and the under-carriage, when passing over rough ground. Inv. 1930–2.

155. FOUR-WHEELED CART. (Scale 1 : 4.) Lent by The Royal Scottish Museum.

This is a model of a cart which was built early in the nineteenth century primarily for hauling sacks of grain on the roads.

It is not sufficiently compact and the front wheels have too limited a lock for it to be very useful for field work, although the wagons used to-day on the Sussex Downs are clumsier than the one represented. It can be used with a single shaft horse or a pair hitched in tandem, or the single shafts can be removed and a pair of horses can be hitched abreast. Inv. 1928–417.

156. LORRY. (Scale 1 : 6.) Lent by The Royal Scottish Museum.

This model, made late in the nineteenth century, represents a particularly neat type of lorry suitable for road work and field work.

The lack of sides makes for easy loading, and the front wheels have practically an all-round traverse so that it can manœuvre in a restricted area. The draught bar can be used for a single horse, tandem or pair abreast. Inv. 1928–499.

157. WOOD CART. (Scale 1 : 4.) Lent by The University of Edinburgh.

This is a model of a strongly built cart used for hauling firewood in the early nineteenth century.

It would also be used for other rough work about the farm and may indicate something of the earlier form of the Scotch cart. Inv. 1928–622.

158. TIMBER CART. (Scale 1 : 4.) Lent by The Royal Scottish Museum.

This is a model of a type of wagon, used for hauling timber, which was popular in parts of Scotland in the early nineteenth century.

The frame is strongly built and the wheels are made as small as is compatible with rough ground and easy draught, so that in loading long timber over the side of the wagon, it would not be necessary to lift the load through too great a height. The sides are carried up clear of the wheels, so that when round timber is rolled over the side it may clear the wheels. Inv. 1928–418.

159. TIMBER WAGON. (Scale 1 : 4.) Lent by The Royal Scottish Museum.

This is a model of an apparatus invented by Sir Henry Stewart prior to the year 1831, for the purpose of transplanting large trees.

It was wheeled up to the growing tree, the shaft was raised to a vertical position and was lashed to the tree trunk some distance from the ground, and the lower part of the trunk was made fast to the " Pillow " between the wheels. The roots of the tree were freed by digging around them and cutting underneath, the wheels were blocked and the tree was drawn out of the ground by hauling on the pole until it assumed a horizontal position. Horses were attached to the two hooks on the axle and the tree was then moved to its new position and replanted by reversing the above process.

Inv. 1928–514.

160. WATER CART. (Scale 1 : 4.) Lent by The University of Edinburgh.

This is the model of the common cart that is used for conveying water to stock which have no access to ponds or streams, and for carting water to the engines during threshing and steam ploughing.

It comprises a barrel to hold about 50 gallons of water, and wheels, axle, frame and shafts. For the most part there has been no attempt to improve on the farm water cart.

Inv. 1928–620.

161. THE DAWE-WAVE WHEEL. Lent by J. Dawe, Esq.

This wheel illustrates an attempt on the part of Mr. J. Dawe to solve the problem of the gripping of the soil surface by a tractor wheel.

The paraffin-burning tractor which to-day is used on the farm for ploughing, harrowing, cutting corn, etc., is really an adaptation of the motor-car, and in the process of adaptation not the least important of the problems which arise, is the designing of a driving wheel which will take hold of the soil, and make the best use of the power of the engine on any surface. A narrow flat rim will sink into the soil too deeply ; a broad flat rim would allow the wheel to slip round without moving the tractor forward ; round spuds of any length waste the power of the engine in stirring the soil unnecessarily and tend to allow the wheel to slip in loose or wet soil ; strakes or broad spuds set across the rim tend to clog even in comparatively dry soil, and the orthodox broad rim with strakes or spuds leaves a track where increased disturbances and compression of the soil has occurred, a feature which certainly results in uneven treatment of the soil, and which is even said to prevent fertility. The Dawe-wave wheel is designed to overcome these difficulties by making narrow rims which are waved in the form of a simple harmonic curve, and which are attached to the spokes to form the tread of the wheel. In improved forms of the wheel the rims are much narrower and a slightly different arrangement of the waves in the rim has been adopted.

The improved wheel is said to allow the tractor to run on the road without any harm to the surface, and to allow the wheel to sink sufficiently in loose soil to obtain a reasonable grip, while in damp sticky soil, although the wheel has a firm grip, there is stated to be practically no tendency for the spaces between the waved rims, to clog with earth.

Inv. 1929–817.

162. OX YOKE. Copied from an original in the Torquay Museum.

This is an example of an ox yoke as used in Devon about the middle of the nineteenth century.

The horizontal piece of timber lay across the necks just in front of the withers of a pair of oxen, and was kept in position by the two bows each of which looped round the neck of a bullock. The beam of the plough, or the shaft of a wagon, or a rope, passed between the two oxen and was fastened to the middle of the yoke. Comparison with a horse's harness shows that whereas the bullock pulled by pressing with its withers, a horse applies its weight at the shoulder joint, a point about two-thirds of the length from the top of the collar.

Inv. 1928–289.

163. CART HARNESS. (Scale 1 : 4.) Lent by The Royal Scottish Museum.

The four types of model harness shown on these horses is typical of that used on farms throughout Great Britain. It will be seen that there is no very great difference except in detail, between the four types shown.

Inv. 1928–398.

BARN MACHINERY

164. ROLLER MILL. (Scale 1 : 4.) Lent by The Royal Scottish Museum.

This is a model of one of the earliest kinds of machines for preparing grain for feeding to stock, and was introduced about the year 1850.

It consists of a pair of heavy iron rollers, mounted in a stoutly built wooden frame. The grain was fed from the hopper into the tray above the roller housing, it then passed between the rollers, and the delivery spout probably shot the meal through a hole in the floor into a bin on the floor below. The belt pulley is attached to the shaft of the fixed roller, and on either side of the roller there is a cam which works against each of the two legs of the feed tray, thus giving the tray a rocking motion. The adjustable roller is driven by gear wheels from the shaft of the fixed roller, its two bearings being mounted in slides, so that by turning the two screws at the front of the machine, this roller can be moved up to any desired distance from the fixed roller.

Inv. 1928–361.

165. ROLLER MILL. (Scale 1 : 4.) Lent by The Royal Scottish Museum.

This mill is a product of the mid-nineteenth century and is an example of a rather later type of roller mill which was built in the barn.

It was used for bruising oats or barley for cattle food. One roller is driven direct from the belt pulley and the other is driven by the first, through a pair of gear wheels. The driven roller is mounted in sliding bearings so that by manipulation of the two screws which react on the bearings, adjustment of the distance between the rollers can be effected. The grain falls from the hopper on to the rollers, the rate of feeding being adjusted by means of a counter-weighted lever which operates a trap in the bottom of the hopper.

Inv. 1928–533.

166. GRINDING MILL. (Scale 1 : 4.) Lent by The Royal Scottish Museum.

This is an example of a nineteenth century mill for producing barley meal or for grinding wheat or oats.

In this model the left-hand mill has been stripped of its housing and the upper stone lifted aside, to give some idea of the appearance of the surface of a freshly dressed stone and to show how the upper stone—the " Runner " in this case—fits on to the spindle. The fineness of the meal which results from grinding depends on how close together the stones can be set. The adjustment of the stones is effected in this model by raising or lowering the vertical spindle together with the upper stone. A lever runs from its fulcrum under the mill, passes below the end of the spindle and is attached to a vertical rod, the top end of which is fitted to a capstan. The rotation of the capstan moves the end of the lever in a vertical arc, and so raises or lowers the spindle and the stone. The upper millstone is driven by the rotation of the spindle which receives its motion through bevel gear wheels from the horizontal shaft. The gear wheel on the spindle can slide up and down and the mill can be stopped quickly by means of the brass handle, which, acting through a quadrant, can raise the spindle gear wheel clear of its driver. The grain runs from the hopper through the hole in the centre of the upper stone and falls on to the bed stone ; thence it works its way between the stones and falls clear at the circumference into the mill casing, passing into a spout which carries the meal to the floor below. Inv. 1928–534.

167. CAKE BREAKER. (Scale 1 : 4). Lent by The Royal Scottish Museum.

This model was submitted to the Highland Society by the inventor in 1841, and it represents a machine for breaking up oilcake into small pieces.

In all cake breakers toothed rollers driven at a slow speed are employed, and in this case the rollers are driven through a reducing gear from the shaft which carries the handle and the fly-wheel. Above the rollers is a wooden box which is made to such dimensions that a slab of oilcake is held in it loosely and can sink of its own weight into the rotating rollers.

Inv. 1928–365.

168. OILCAKE CRUSHER. (Scale 1 : 4.) Lent by Messrs. R. Garrett and Sons.

This model, shown at the 1851 Exhibition, represents a machine introduced in 1841 for breaking up oilcake into small pieces suitable for feeding cattle, or for re-ducing it to a powder for use as manure. It is driven by a winch handle attached to a fly-wheel that drives the lower pair of rollers which are fluted, and by reducing gear also drives the upper rollers which have intersecting teeth. The distance between the rollers is in each case adjustable by sliding bearings, and the spaces between the cogs of the upper rollers are cleared by fixed scrapers. Inv. 1894–166.

169. CHAFF-CUTTER. (Scale 1 : 4.) Lent by The Royal Scottish Museum.

This is a model of an early type of chaff-cutter, similar machines being in use in England before 1830.

The straw is fed into the box and forced against the two rollers, which grip the straw and push it down the funnel where it becomes tightly packed. As it emerges from the funnel it is caught by the knives, which have partly a chopping action and partly a scissors action against the edges of the funnel. The driving shaft carries the handle, the two knives, and a fly-wheel ; the second shaft, driven much more slowly than the driving shaft through a pair of gear wheels, carries the bottom roller and drives the shaft carrying the top roller. Inv. 1928–363.

170. CANADIAN STRAW-CUTTER. (Scale 1 : 4.) Lent by The Royal Scottish Museum.

This machine for preparing straw for feeding cattle was brought into Scotland in 1837.

The first shaft carries the handle, a small driver gear wheel and the fly-wheel, while a second shaft, driven by the first, carries a roller provided with longitudinally set knives. The straw which is fed into the box is caught by the knife roller and cut against a wooden idler roller set below it. Inv. 1928–362.

171. CHAFF-CUTTER. (Scale 1 : 6.) Presented by J. M. Taylor, Esq.

This is a model of a hand-power chaff-cutter typical of those introduced in the middle of the nineteenth century.

The main shaft carries the fly-wheel with its adjustable knives, and a worm for driving the feed rollers. The forage is placed in the box and is pressed forwards against the rollers ; these seize and compress it and present it to the knives which cut it scissors fashion against the steel sill of the box.

This machine differs from earlier types in having a better designed cutting mechanism. The sill is armed with a steel bar which has a right-angled edge against which the knife slides, and the knives are provided with set screws so that their edges can be made to meet the sill. It differs from later models in having no travelling apron in the box to assist in drawing the forage forward, and in having a fixed upper feed roller. Later designs provide an upper feed roller which has vertical play against springs or a weighted lever. Inv. 1929–1047.

172. CHAFF-CUTTER.

This is the cutting mechanism of a chaff-cutter which was found in Somerset, and which was probably made late in the nineteenth century.

The straw or hay is gripped between the two rollers, the lower one being fixed and the upper one being held down by a flat steel spring. As the rollers rotate the compressed straw is forced forward into a rectangular slot, formed by an iron casting. Three steel blades are mounted on two cast-iron discs to form the drum of the chaff-cutter. The blades work against the lower edge of the slot on the side away from the rollers, so that as the straw is forced through the slot, it is cut by a scissors action between the rotating blade and the stationary sill.

This type of cutting mechanism is not common on English chaff-cutters, but it has been re-introduced into this country recently in some types of silage cutter and blower. Inv. 1928–195.

173. CHAFF-CUTTER. (Scale 1 : 4.) Lent by Messrs. R. Garrett and Sons.

This model was shown at the 1851 Exhibition, and represents the ordinary form of machine by which forage for horses is cut into short pieces resembling chaff. This type of machine was originally patented by James Cooke in 1794.

It consists of a horizontal trough for receiving the uncut forage, which is moved by hand towards a pair of fluted feed rollers that force it through the mouth of the machine. Here two revolving knives attached to a fly-wheel cut off the material fed through twice every revolution. Pressure is applied to keep the material compact at the mouth where it is being cut, the upper roller and a sliding block being continually pressed downwards by a lever and weight arranged under the trough. The fly-wheel is directly rotated by a winch handle, but additional power can be applied by another winch which is connected with the former by bevel gearing and a clutch. The feed motion is derived from the fly-wheel shaft by spur and bevel gearing, and two rates of speed can be obtained by the use of a sliding pinion. The knives are curved and they cut with their concave edges, but they are usually made to cut with their convex edges, as in the original machine. Some machines are fitted with an intermittent feed motion which holds the forage stationary while being cut. Inv. 1894–165.

174. CHAFF-CUTTER AND SIFTER. (Scale 1 : 6.) Presented by J. H. Turner, Esq.

This model shows a power-driven combined chaff-cutter and sifter containing improvements patented by Messrs. J. H. Turner and G. A. Gray in 1904.

The forage to be cut is fed into an upper trough and is carried by travelling crossbars towards the end. It is there compressed and forced through the mouth of the trough by a fluted roller and a pair of toothed rollers, the upper one of which is pressed downward by a weighted lever below. Four curved knives attached to a revolving wheel cut off the chaff which then falls into a wooden casing leading to the upper one of a pair of sieves. These sieves are inclined and placed one above the other, and their lower ends receive a small circular motion from opposite eccentrics, mounted on a shaft above them, while their upper ends have a longitudinal motion only. The upper sieve has wooden screens which prevent the " cavings " or long chaff from passing through and carry it to an elevator that returns it to the feeding trough. The short chaff falls through on to the lower sieve, which is fitted with metal screens for removing dust and dirt, and passes to a spout at the end. The feeding rollers and travelling bars are driven by toothed gearing from a longitudinal shaft, to the end of which the knife wheel is fixed, through a bevel wheel reversing gear. If the machine should choke, this gear is automatically reversed by a roller which rests on the top of the forage as it is being fed in. The gear might also be actuated by a person who, by any means, got caught in the machine. Inv. 1910–24.

175. ROOT-CUTTER. Lent by The Royal Scottish Museum.

This implement is used for chopping up mangolds and turnips to a convenient size for feeding cattle.

The edges on the under-sides of the blades are kept fairly sharp and the implement itself is moderately heavy, so that when used with both hands on a wooden floor or a chopping board it is a quite effective but rather slow method. Inv. 1928–364.

176. KILN. (Scale 1 : 4.) Lent by The Royal Scottish Museum.

This model, which represents an invention introduced during the latter half of the nineteenth century, shows in skeleton form the arrangement of a kiln for drying wheat preparatory to milling.

Many of the wheats which are used in the manufacture of flour, particularly those which are grown in hotter countries, do not need to be passed through the kiln, but home-grown wheat frequently has too high a percentage of moisture and it is often an advantage to toughen the seed-coat, thus producing a coarser and more easily separated bran. The model shows the metal frame-work of a building having a furnace on the ground floor. The floor above the furnace room comprises perforated metal sheeting laid on metal joists. The grain is spread in a layer on this floor, while the heated air rises from below assisted by ventilating shafts in the roof and passes up through the layer of grain, raising its temperature and robbing it of its moisture. The air temperature is modified by control of the fire in the furnace by means of the damper shown at the end of the model. Inv. 1928–528.

177. RUSSIAN DRYING HOUSE. (Scale 1 : 4.) Lent by The University of Edinburgh.

This is a model of the kind of kiln used in Russia in the eighteenth and nineteenth centuries, for drying grain.

Kiln drying of grain in the straw was found necessary in Russia, Livonia, Courland and Lithuania, and the practice was extended to such crops as peas, beans and buckwheat. Drying is allowed to proceed as far as possible in the field and is completed in the kiln. The crop is ready to be threshed as soon as it is sufficiently dry, but as a rule oats and barley were threshed during the winter by degrees, as they were needed by the stock.

The interior of the model is divided into three roughly square-shaped main compartments, the two outside compartments being the drying rooms, and the middle one being the threshing floor. Each drying room contains a stove on the ground floor and above it two floors formed of joists, on which the crop is laid. The wood is charred before it is actually burned, and the stove is constructed so that there shall be no danger from sparks. The long corridor outside the three large compartments served as a clearing house for bringing corn to, or taking it from, the kiln. Inv. 1928–637.

178. CORN SCREEN. (Scale 1 4.) Lent by R. Boby, Esq.

This machine was patented by Mr. Boby in 1855, but some of the details are later improvements ; it is used for separating thin grain and dust from corn, so as to obtain a uniform product.

The machine consists of a sieve, sliding on guides inclined at an angle of 13 deg., upon which it is reciprocated by a rapidly rotated crank shaft. The corn, from a hopper above, is passed through a distributing box that delivers it at the top of the screen, with the result that the smaller seeds drop through the meshes of the sieve into a receptacle below, while the larger ones pass off at the foot. The screen itself is detachable so that the grade can be altered ; that in the model is of an actual mesh, so that this model machine is suitable for use in testing samples of grain.

To prevent the screen from being choked, there are, fixed below it to the stationary framing, transverse rods upon which are strung loose washers that project up between the wires of the screen, so that when the screen is in motion these washers keep the intervening spaces clear. The wire rods of the screen are of rectangular section, but have rounded top edges to assist the smaller grain into the spaces. Inv. 1898–55.

179. SCREEN WITH BLOWER. Lent by Messrs. Robert Boby, Ltd.

This machine is used by horticulturists for treating small parcels of seed and by seed merchants for testing samples from bulk grain.

Before any sample of seed, whether it be wheat, barley, oats, clover, grass seeds, etc., can be sold to the grower, it is essential that it shall contain no weed seeds, broken grains, dirt, or bits of straw. Also, if it be seed wheat, there must be no barley or anything else but wheat present, and the individual grains should be of an even size. To ensure that only pure, clean, even samples of seed are sent out, seed merchants make use of various cleaning machines.

The example illustrates one type of the machines used. One man turns the handle while another pours the seed to be cleaned into the hopper of the machine, bags up the finished product and deals with the rubbish and by-products. The turning of the handle causes a rocking movement of the screens and rotates the fan which is inside the cylindrical housing below the screens.

The seed in the hopper falls on to the first screen, passes through it and on to the second screen, through which it also passes and is held by the third screen. Bits of straw and other seeds which are larger than the pure sample are held by the first two screens, and discharged by the shute provided, to the side of the machine opposite the handle. Dirt and other seeds which are smaller than the pure sample, fall through the third screen and are discharged at the side of the machine near the handle. By this means material and seeds which differ in size from the pure sample are separated. In the next process, seeds, etc., are separated which are of the same size, but of lighter weight than the pure sample.

All that is retained by the third sieve is shaken into a tray which allows the seed to fall into the tunnel which leads away from the fan. It is then obliged to fall down this funnel against the draught, the strength of the blast being adjusted so that the heavy even sample can fall into the box at the bottom of the machine, while lighter material is carried on up the tunnel and is blown clear. Inv. 1928–31.

180. POTATO WASHER. (Scale 1 : 4.) Lent by The Royal Scottish Museum.

This is a model of an apparatus introduced in the mid-nineteenth century, and still in use for washing potatoes in preparation for use in the house or for marketing.

It consists of a cylindrical wooden cage, provided with a door through which potatoes can be fed or emptied, and mounted on a shaft so that it can be rotated. Hung so that it can swing freely from the same shaft, is a trough so shaped that the lower part of the cylinder is enclosed by it, and provided with a lever for tilting it, so that it can be swung through 45 degrees and emptied. When the trough is filled with water and the cylinder is rotated, potatoes contained in the cylinder are rubbed against each other and against the bars, and the dirt is taken up by the water. When the potatoes have had sufficient treatment the trough is swung up out of the way and held up by engaging the lever in a catch on the frame. The door in the cylinder is then opened and turned downwards, allowing the potatoes to fall clear of the apparatus. Inv. 1928–366.

181. BATH STOOL. (Scale 1 : 4.) Lent by The Royal Scottish Museum.

This is a model of a stool used in the eighteenth and nineteenth centuries in the operation of " bathing " or " pouring " sheep.

Pouring was carried out as follows :—The shepherd sat astride the narrow end of this stool with the sheep placed on the rack with its head towards him and its legs hanging through the bars. As the shepherd parted the wool with his hands, an attendant boy or woman poured lotion from a jug, on to the fleece, between his hands. When they had worked along its back, the sheep was turned first on one side then on the other, and finally was turned on its back so that the whole fleece was covered.

Inv. 1928–423.

182. SHEEP DIPPER. (Scale 1 : 4.) Lent by The Royal Scottish Museum.

This pattern of sheep dipper was introduced in the middle of the nineteenth century in the early days of the practice of dipping.

In this dipper the sheep is lifted by two men, and dropped into the bath, containing disinfectant fluid. It is then placed on the sloping rack on its back and the disinfectant is worked by hand well into the fleece. Inv. 1928–448.

183. SHEEP SHEARS. Lent by The Royal Scottish Museum.

These shears illustrate the pattern that is very widely used to-day ; in fact it is almost the only pattern of hand sheep shears. Inv. 1928–436.

184. CLIPPERS. Lent by The Royal Scottish Museum.

This mid-nineteenth century pair of clippers is an example of the type introduced for clipping horses. Inv. 1928–563.

185. SHEEP SHEARING MACHINE. Lent by Messrs. R. A. Lister and Co., Ltd.

This exhibit is the hand-piece of a modern sheep shearing machine sectioned to show how the rotary motion of the driving mechanism is transformed into the reciprocating motion of the knife.

The hand-piece carries at one end the clipping mechanism, which comprises a comb, and a knife which lies on the comb. The knife is carried by a large lever which is pivoted at its end farthest from the comb. A short shaft runs through the hand-piece and carries a small disc at one end and a bevel gear wheel at the other end, which engages with another gear wheel, forming a hinged joint, the second wheel being carried on the end of a short enclosed shaft. This shaft receives rotary motion from the driving mechanism through a flexible shaft, the combination of flexible shaft and hinged joint giving complete freedom of movement at the hand-piece.

The rotation of the short shaft is communicated through the bevel gear wheels and the hand-piece shaft to the disc in the forward end of the hand-piece. This disc carries on its forward face an eccentrically-mounted boss which fits into a slot at about the middle of the lever which carries the knife. Since this lever is pivoted at one end, the rotary motion of the boss in the slot causes the other end of the lever which carries the knife, to move from side to side. The fingers of the knife and the teeth of the comb are so fashioned that as the knife reciprocates on the comb the steel edges of the fingers slide along the steel edges of the teeth and produce a somewhat similar action to that of the blades of a pair of scissors. The maintenance of an efficient cutting mechanism consists of preventing by grinding any tendency for the cutting edges to become rounded, and of maintaining by means of the thumb-screw to the front end on top of the hand-piece, sufficient pressure of the knife on the comb.

Inv. 1928–68.

MILLING MACHINERY

186. ROLLER FLOUR MILL. (Scale 1 : 24.) Presented by Messrs. Thos. Robinson and Sons.

The building of the complete roller mill represented in the model is divided into two sections by a vertical fire-proof wall, which separates the wheat-cleaning department, in which a dusty and somewhat inflammable atmosphere exists, from the flour milling department, in which the reduction takes place.

The grain as unloaded into the cleaning department, is delivered into a hopper, from which it is raised vertically into storage bins at the top of the building. Under each bin is a mixing conveyor, which draws off the wheat in any required proportion, and elevates it to the cleaning machines, in which it is automatically freed from impurities and is scoured, washed, dried, and thoroughly cleaned. It is then delivered by a shoot through the partition wall into the cleaned wheat bin.

In the flour milling department the grain after passing through the roller mills is conveyed to the flour dressing and purifying machines, by which the products are classified during the process of gradual reduction, the fine flour for delivery into certain sacks, while bran, offals, and germs are delivered into other sacks.

The mill represented is capable of producing about two sacks or 560 lbs. of flour per hour when driven by an engine indicating 20 h.p. This power is applied to a lay shaft in the basement, from whence it is transmitted by belting and some shafting to the entire plant.   Inv. 1903–5.

187. SEED CLEANER. (Scale 1 : 4.) Lent by The Royal Scottish Museum.

This model illustrates the kind of cleaning machinery which was used in the flour milling industry in the early half of the nineteenth century.

The machine represented, removed from wheat dirt and dust which adheres to the seed-coat, and which cannot be separated by such processes as sifting or blowing. The grain runs from the hopper into the hollow cylinder, which is rotated on its longitudinal axis by means of a bevel gearing driven from the handle at the side. The inner surface of the cylinder is roughened, and while the wheat is passing through it the grains are rubbed against each other and scrubbed by the surface of the cylinder. Dust adheres to this rough surface and is carried up to be removed by the brush fixed inside the cylinders and above the axle. The dirt falls into the trough in which the brush is fixed and is carried out to a spout, while the grain falls into the channel below and is conducted to a separate spout at the side.   Inv. 1928–360.

188. BRUSH SMUT MACHINE. (Scale 1 : 4.) Lent by The Royal Scottish Museum.

This model shows in section a machine which was brought out in the latter half of the nineteenth century for removing fungi, dust, and sand from the seed-coat of wheat before milling.

A vertical shaft carrying spirally arranged brushes is enclosed by a wire gauze screen so that the brushes are in actual contact with the screen, the whole being enclosed by a dust-proof casing. Grain is run from the hopper into the inner chamber and is scrubbed by the brushes as the spindle revolves. The cleaned grain cannot pass through the small mesh of the screen but runs out at the bottom of the machine from the central spout. The dust, etc., is forced through the screen by the combined effects of the brushes, centrifugal force, and the air currents set up, and passes out of the machine by the second spout.   Inv. 1928–529.

189. CHILDS' ASPIRATOR. (Scale 1 : 2.) Presented by B. Corcoran, Esq.

This machine, for cleaning grain and separating from it all dirt and foreign seeds, was patented by Mr. A. B. Childs in 1852, 1864 and 1867 ; it was, for about 30 years, the principal machine used in flour mills for cleaning the wheat before grinding.

The apparatus consists of a closed wooden box having a revolving fan mounted at one end of it, a grain-receiving hopper at the other end, and a double screen mounted above it. The grain is fed on a dead plate at the upper end of the screen, and is caused by a baffle-plate to spread out into a thin layer. The screen is mounted on four wooden pillars ; it receives a reciprocating motion from an eccentric on a transverse countershaft and also receives a vertical blow from hammers below it at the end of the forward stroke. The screen has two perforated plates, one above the other ; the upper one has holes, through which the wheat falls, while larger objects pass over to a spout at the end ; the lower plate allows the small seeds and dirt to pass through it to another spout, and delivers the wheat into the hopper. One side of the latter is adjustable for controlling the feed and the bottom is formed by a slowly rotating grooved roller which, by the aid of leather strips on the hopper, allows the wheat to pass without admitting air. The wheat passes downwards to an opening near the lower end of a narrow vertical passage, through which a current of air is drawn by the fan ; this air-current is so adjusted that the wheat falls to the bottom of the passage, while all materials of less density are sucked up into a large chamber, where the chaff and husks are deposited, while the light dust is blown out through the fan opening. An adjustable regulating valve is placed in the top of the chamber to keep the air-current constant in spite of variations in the speed of the fan ; the chamber discharges its contents automatically through a bottom door which is closed by a spring.

The aspirator represented would deal with 25 bushels per hour, the speed of the fan being 400 revs. per min. Inv. 1911–4.

190. MILL. (Scale 1 : 4.) Lent by The Royal Scottish Museum.

This is a model of a type of mill that is used in Ceylon. It is interesting because it is made on the principle of the pestle and mortar, the principle which was probably exploited first for the purpose of crushing grain to make flour. Inv. 1928–553.

191. QUERN OR CORN MILL.

This hand appliance illustrates the means used from very primitive times for grinding corn. The mill shown came from the island of Barra, in the Outer Hebrides, where such mills are still occasionally used ; they were quite common within living memory.

The mill comprises a nether millstone or bedstone, which is fixed, and an upper millstone or runner, which is turned by a handle. The section of the stones is slightly conoidal. There are no furrows in the faces, but the runner has an irregular surface which seems to act in a similar way. In a hole in the centre of the bedstone a wooden peg or spindle is wedged. Across the " eye " in the centre of the runner is wedged the rynd or crosspiece, which has a bearing on the underside to run on the spindle. Another hole in the runner accommodates the handle by which it is turned. Corn is fed into the eye, and is ground finer and finer as it approaches the circumference, where it is delivered as meal.

The stones are of granite and about 21 in. diam. Inv. 1912–68.

192. ESSEX MILL.

This type of mill was used for grinding corn by hand in the eighteenth century, and the example shown came from Fyfield Hall, Ongar, Essex.

It is a particularly good specimen of millwright's work as practised in the eighteenth century and earlier showing how, before the use of cast iron became common, the difficulty of making mechanisms such as bevel gearing in timber was overcome.

It illustrates a step in the development of the flour mill and represents an early form of the mechanical mill. The circular mill with the upper stone which is turned with a stick fitted into a slot is quite a primitive implement. In principle this mill is much the same except that the upper stone is mounted on a spindle which pierces the bed stone and which is turned by hand through a bevel gearing. The workmanship involved in fashioning the stones is, of course, far from a primitive standard.

Inv. 1913–491.

193. CONICAL FLOUR-MILL. (Scale 1 : 4.) Presented by H.R.H. Prince Consort.

This model represents a flour-mill patented by Mr. Walter Westrup in 1850. It has two pairs of stones arranged on a single vertical shaft : in each set the lower stone is the runner and is somewhat conical with the smaller end uppermost, while the bedstone is annular and bevelled to fit. Between the two mills is a cylindrical drum, with gauze screens acting as a dresser, into which the meal from the upper mill is delivered, the flour passing off while the coarser meal falls down into the lower mill where the grinding is finished.

In the actual mill the runners were 30 in. diam., and are stated to have been capable of grinding twice as much flour as a pair of ordinary 4 ft. stones, while only 0·1 of their weight. This successive grinding with intermediate screening is the leading feature of modern milling ; even with this two-stage mill it was found that the loss by middlings was reduced by this treatment. Inv. 1857–103.

83

194. DRESSING ENGINES. (Scale 1 : 4.)   Lent by The Royal Scottish Museum.

These are models of late-nineteenth century machines for separating flour into various grades.

In both machines use is made of the principle of the rotary screen.   In one model the screen is covered with silk, so that only the finest flour passes through it to fall into the trough below.   Here a screw conveyor carries the flour along and discharges it at the end of the dresser ; the coarser material is dropped into a wooden casing at the end of the screen and runs to a spout set to one side of the flour exit.   The screen in the second model is covered with three different meshes of wire gauze, each section of the screen having below it a separate compartment leading to a separate spout. Three different grades of flour are collected and the bran is discharged at the end of the screen.   Inv. 1928–535, 536.

195. CYCLONE DUST-COLLECTOR. (Scale 1 : 4.)   Lent by Henry Simon, Esq.

This apparatus, introduced in 1867, is for removing and collecting the dust raised in grinding, flour-milling, and similar operations.   It consists of a short vertical cylinder with a tangential inlet, and a central outlet tube at the top, while the lower end is formed by an inverted cone with a small outlet at its apex, all made in sheet iron.   The dust-laden air is forced in at a high velocity by a centrifugal fan, and, by its tangential entrance establishes a vortex in which the velocity of the air increases, so that the dust particles on account of their density are forced by centrifugal action against the sides of the cone where their energy is absorbed by friction.   They then slide down to the apex of the cone by gravity, where they are collected ; spiral ribs fixed inside the case assist this downward movement.   The air, freed of its solid particles, rises in the centre, and escapes upwards through the tube.   Inv. 1890–88.

196. RICE MILLS. (Scale 1 : 8.)   Lent by The Royal Scottish Museum.

These are two models of nineteenth century wooden mills used by the Chinese for grinding rice.

Both models show the two " stones," the upper being the runner and the lower the bedstone, while the more complete model shows how the runner is mounted and how the mill is built up on a four-legged wooden platform.   The two pieces of timber are faced on their grinding surfaces in a manner which is similar to that found on our own mill-stones.   The rice is fed into the mill at the centre and the meal is delivered at the circumference of the grinding surfaces.   Inv. 1928–552, 609.

197. RICE MILL.   Lent by The Royal Scottish Museum.

This is a nineteenth century Japanese hand rice mill, made of blue glazed porcelain.

The upper stone, which is the runner, is of a considerable height to give it the necessary weight, and it is provided with a hollow centre through which the rice is fed.   The square holes in the sides are the sockets of four handles by which the mill is turned.   The bed stone is expanded into a trough which catches the meal and the two grinding surfaces are faced in the usual way.   Inv. 1928–546.

198. RICE MILL. (Scale 1 : 6.)   Lent by Messrs. Douglas and Grant, Ltd.

This model represents a standard " Jungle " type mill to treat paddy or rice in the husk.   The main processes consist of cleaning, removing the husk, removing the small, broken or dead grains, removing the cuticle, and whitening and polishing the grain.

The paddy is first passed over a shaker or sift which removes sticks, straws, stones, etc., and the cleaned paddy is raised by an elevator and discharged into a huller or sheller which removes the husk from most of the grain.

The rice and husk fall into a second elevator (the model shows two in one casing) and are fed over a second shaker to remove small, broken or dead grains as well as split husk.

The cleaned rice and husk pass directly to a winnowing fan, which blows out the husk and dust, leaving behind the full grain, mostly shelled, but still containing some smaller and unshelled grain.

A further separation now takes place, the still unshelled grain being returned to the huller, whilst the shelled grain passes to the first of two " white rice " cones.   In this cone the cuticle, which is sometimes dirty and of a deep red colour, is removed by friction and the grain is whitened.   The second cone shown is to further whiten the rice and in large installations the rice is subsequently burnished.

From the second cone the rice passes to a sift and aspirator which grades the rice into whole rice, large broken, and small broken, and also removes the particles of free meal, and small points of rice. The offals, meal, points, etc., are separately collected as marketable products.

Power to drive the mill is supplied by a steam engine, fuel for the boiler consisting, in all but small installations, of the husks collected by the mill.

The model represents a mill to treat about one ton of paddy per hour.

Inv. 1921–601.

199. BARLEY MILLS. (Scale 1 : 4.) Lent by The Royal Scottish Museum.

These two models illustrate the type of mill which is used in preparing barley for domestic use, the primary object in milling barley for this purpose being to remove the husk.

One model shows the complete mill with the gearing, while the other has been stripped of most of its parts to show the stone. In the complete mill the main shaft carries the main belt pulley, a small pulley for driving the indicator worm shaft, a large gear wheel for driving the shaft which carries the stone, and a small gear wheel which drives the casing. It will be noticed that the casing revolves at a considerably lower speed than the stone. A definite quantity of grain is fed into the casing through the openings on the rim which are then closed, and the grain is rubbed between the stone and the casing to a definite extent indicated by the number of revolutions shown on the dial. The treated barley is then taken from the mill, the husks are removed, and the meal is graded into pot barley and pearl barley. Inv. 1928–530, 531.

200. BARLEY SIZER. (Scale 1 : 4.) Lent by The Royal Scottish Museum.

This is a model of a machine used in the nineteenth century for grading barley into pot barley and pearl barley.

In addition to its use in brewing, barley is also used in domestic cookery, and for the latter purpose the harder types of the grain are preferred. To make it suitable for use in the kitchen, it is necessary to remove, by grinding, the hard cuticle, and when, after grinding, the mass is graded, the larger grains are known as pot barley and the smaller pellets as pearl barley. Patent barleys are usually the whole mass ground to a flour.

The machine comprises a rotary screen on which it is possible to use gauzes of various meshes, the smallest mesh being nearest to the feeding end of the screen. Beneath the screen are four spouted compartments and a fifth at the end of the screen, so that by using gauzes of four different meshes it is possible to divide the bulk into five fractions. The brush which lies against the screen, is for the purpose of forcing back into the interior, any material that may lodge in the gauze.

The mass of barley pellets and flour is introduced from the hopper at the upper end of the rotating screen, the finer products make their way through the mesh and into the various compartments, while the pot barley is discharged at the lower end of the screen. Inv. 1928–532.

201. CORN-WEIGHING MACHINE. (Scale 1 : 4.) Lent by W. H. Baxter, Esq.

This is a model of a self-acting and self-registering weighing machine for corn, by Baxter.

Corn falls from a fixed hopper, through an intermediate hopper, into a receptacle balanced by a weight, and capable of turning on its axis, when it descends and a catch is let go. There are four such receptacles, forming a drum, and when sufficient corn has passed into one receptacle to overbalance the weighted lever, the dropping of the former sets free the catch and allows the receptacle to turn round and tip out the corn. Each motion of the balance is recorded by a counter. On the descent of the receptacle the mouth of the hopper is closed. Inv. 1874–80.

DAIRY MACHINERY

202. MILKING PAIL. (Scale 1 : 4.) Lent by The University of Edinburgh.

This is a model of a Scottish nineteenth century milking pail of traditional pattern. This form of open shallow pail made of timber was once used over most parts of these islands, but the type has had to give way to the less picturesque but more practical and hygienic bucket. The timber bucket would be liable to deteriorate under alternate wetting and drying, and the wood itself would harbour dirt and bacteria. The wide shallow shape has also fallen into disfavour as it presents a proportionately larger surface area on which dust, dirt and bacteria can collect. Inv. 1928–635.

203. MILKING MACHINE. Lent by Messrs. The Alfa-Laval Co., Ltd.

This machine, introduced recently for milking cows, is the development of an invention by Mr. Daysh, of New Zealand.

This machine, in common with most modern ones, applies four rubber-lined metal cups to the cow's teats. In this case the rubber lining is in the form of a tube stretched lengthways inside the metal cup so that the space between the rubber and the metal forms an annular airtight chamber. When the teat is fitted into the rubber tube, variations in the air pressure in the rubber tube below the teat, and in the annular space cause the application of suction to the teat, and alternate pressure and release of pressure of the rubber tube on the teat, processes which imitate fairly closely the action of a calf's mouth.

The space inside the rubber tube is connected to the receiving bucket by a pipe along which the milk flows, and the bucket is connected direct to the main vacuum system in which low pressure is produced by means of an engine-driven air pump. In this way, a steady suction would be applied to the teat. The annular spaces of the four teat cups are connected by smaller tubes issuing from the sides of the cups to a metal chamber which receives from the bucket two small-diameter tubes. One of these tubes connects with the main vacuum system and enters the junction chamber at the end nearer the bucket, the other tube enters the chamber at the opposite end, and is connected with the main pulsator which is situated at the side of the main air pump. By this means a piston inside the junction chamber is made to reciprocate, thus interrupting the suction which is transmitted to the annular spaces in the teat cups, momentarily allowing the air pressure to rise in these chambers first in the fore teats then in the hind teats.

When the reduced pressure is the same in the rubber tube and in the annular space, the teat is subjected to suction. When the pressure rises in the annular space the rubber tube tends to collapse inwards and exert pressure on the teat, this mechanical pressure being applied first at the top of the teat and progressing down its length until the tube closes below the teat and cuts off the suction. The ensuing fall of air pressure in the annular space then again relieves the mechanical pressure on the teat, this series of operations being repeated about 40 times a minute.

Inv. 1928–1308.

204. CREAM SEPARATOR. Presented by Messrs. The Alfa-Laval Co., Ltd.

This machine was invented by G. de Laval for separating milk or any other mixture of liquids of different specific gravities, and it embodies patents taken out in 1878 and 1881. The contemporary example shown is one of the few existing machines of the original design.

The bowl differs from the modern bowl in being a spheroid, not cylindrical or conical, and it does not contain the more modern nest of plates. Above the bowl are two spouted trays, the lower one for the delivery of separated milk and the upper one for cream. The two trays are detached and are lying on the base of the separator. Above the trays was a vessel in which the whole milk was held and from which it was delivered by means of a tap into the hollow tube protruding from the top of the bowl. This tube led down to the bottom of the interior of the bowl and being in the shape of an inverted T it delivered the whole milk to the bowl, near the bottom and as far away from the centre as possible. As the bowl revolved the milk was centrifuged, while the constant stream running in at the bottom caused the milk to overflow into the two trays. That fraction of the milk which was nearer the centre of the bowl (the

86

cream) found its way through the upper annular aperture on the neck of the bowl while the heavier portion (the separated milk) came to the lower aperture and so to the lower tray. Inv. 1928–954.

205. CREAM SEPARATOR. Presented by Messrs. Watson, Laidlaw and Co., Ltd.

The machine shown was patented in 1888–1902 by Messrs. J. Laidlaw and J. W. Macfarlane, and is of a size suitable for working by manual power. The milk to be treated is placed in an attached open reservoir, from which it passes to a stationary cylinder at the top of the machine, fitted with a float which automatically regulates the flow to suit the capacity of the apparatus. From this cylinder the milk passes into a distributing cup, formed in the top of a revolving cylindrical drum, and through small holes in the circumference into the chamber below. The chamber is fitted with a nest of double conical aluminium plates which have small distances between them and slots through the outer cones. The heavier portion, or skim milk, collects around the sides of the chamber, having forced the cream into the central region ; the plates divide the milk into thin layers and increase the rapidity of separation. Two vertical tubes lead the cream to the top of the drum, from which it flies off into an annular channel that leads to the cream outlet ; the skim milk passes round the outer edge of a flat plate at the bottom of the chamber to a central outlet which retains a large body of milk in the drum, but permits discharge as fresh milk is fed in.

The separator shown is for treating 50 gallons of milk per hour, when the handle is turned at the rate of 56 revs. per min., this causing the drum to make 8,400 revs. per min., the total ratio of the gear employed being 1 : 150. The gear is all cut and is of the bevel and spur wheel type ; on account of the great amount of energy accumulated in the revolving masses, the winch handle is connected to its shaft by a form of ratchet drive, which allows the shaft to run forward free of the handle. Owing also to the high working speed the lower end of the central shaft rests on a ball pivot, while the upper bearing is carried in a washer of india-rubber, by which arrangements easy running is ensured. Inv. 1908–28, 29.

206. " NEW LISTER " CREAM SEPARATOR. Lent by Messrs. R. A. Lister and Co., Ltd.

This is a sectional example of a modern machine for separating cream from milk.

The milk is first poured into the hopper at the top of the separator. As soon as the machine is running at its normal speed, the tap at the bottom of the hopper is opened and milk is admitted into the feed tray. A tube from the bottom of this tray passes down, and allows milk to run out of the feed tray straight into the bowl. The quantity of milk running into the bowl is maintained at a uniform rate by the float in the feed tray. When the milk in the tray reaches a certain level, the float is pressed against the tap and the flow of milk from the hopper is checked ; as the level of milk in the tray falls, the float is drawn away from the tap and more milk is admitted.

The bowl is fitted with a central tube, open to the feed tray at the top and open to the inside of the bowl at the bottom, so that the chamber where the milk is set spinning actually fills from the bottom. As the milk spins the cream will find its way toward the central tube, and any sand or dirt will move toward the outside of the bowl. At the top of the bowl are two apertures, one toward the central tube and one a little further from the centre, both opening into the bowl. As more milk is run into the bowl through the central tube the level rises until milk is forced through the two apertures. As, however, one is nearer the centre than the other, the lighter material, cream, will come out of the former and the heavier milk from the latter. The bowl is cone-shaped, so that the aperture near the centre is at a higher level than the one further out, an arrangement which makes it possible to fit two receiving trays one above the other, the upper one to take up and deliver to a vessel the jet of cream issuing from the upper aperture and the lower one to deal with the jet of milk. Inv. 1927–2018.

207. CREAM SEPARATOR. Lent by Messrs. The Wolseley Sheep Shearing Co., Ltd.

This is a sectional example of a modern cream separator, showing the internal arrangement of the bowl and the driving mechanism.

It will be seen that the interior of the bowl is divided up by a series of thin conical plates arranged one above the other, into a number of narrow compartments. This is done in order to minimize frothing while the milk is passing through the bowl, and to ensure that the speed of the bowl is communicated to the milk. This section also shows how the milk is admitted to the central tube of the bowl, and thence into the lower part of the bowl itself. At the upper end of the bowl may be seen the channels which lead to the cream and milk apertures.

A complete bowl is also exhibited. Inv. 1927–1949, 1928–1313.

208. DAIRY FURNITURE. (Scale 1 : 4.) Lent by The Royal Scottish Museum and The University of Edinburgh.

These are models of tables, racks, etc., which are required in the dairy, and represent Scottish practice in dairy furniture design in the early nineteenth century.

Inv. 1928–502, 612, 615, 636.

209. PLUNGE CHURN. (Scale 1 : 4.) Lent by The Royal Scottish Museum.

This is a model of the primitive type of churn.

It consists of a cooper-work container and a plunger, which is a short wooden pole with a perforated piston. These churns were in use up to the end of the nineteenth century, and have now practically disappeared. Inv. 1928–419.

210. CHURN. Presented by Messrs. Woolf and Co.

This is a portable form of atmospheric churn, consisting of a vertical tin cylinder, within which freely slides a disc or piston. The piston is fitted with a tubular rod, terminating in a cross handle, by which it may be reciprocated. A removable cover closes the vessel and prevents splashing. The disc and the rod are extensively perforated, and through the holes jets of air or milk are forced when the apparatus is worked. With a stroke of 6 inches and about 60 strokes per minute the butter is extracted in 10 or 15 minutes. Larger churns of this type are constructed with a crank motion for driving. Inv. 1894–191.

211. CHURN. Lent by B. Samuelson, Esq.

When introduced, this was known as an " atmospheric churn " on account of the separation of the butter being assisted by the large amount of air that the machine forced through the milk.

It consists of a rectangular box fitted with a horizontal axle on which are secured four radial paddles ; the paddles being alternately solid and perforated. The cover of the churn is provided with two handles which are hollow, and so serve also as air passages. Inv. 1860–16.

212. HORSE-DRIVEN BUTTER CHURN. Presented by J. Roads, Esq.

This churn was built in the late eighteenth or early nineteenth century, and was still in position though not in use in 1927, at Broughton Manor Farm, near Brierton, in Buckinghamshire. It was usually driven by a horse, but one instance is recorded in which a tractable bull was employed.

An iron fork hangs from the large gear wheel, in such a way that it can be attached to a horse's saddle. The horse walks round and round in a circle from 13 ft. to 15 ft. in diameter, and so rotates the large wheel ; the movement is transmitted through the two gear wheels on the vertical spindle, to the churn. The driving wheel is 15 ft. in diameter and has 240 teeth. The wheel gearing with this has 40 teeth, the third wheel 48 teeth, and the pinion on the churn 24 teeth. The churn itself is upwards of 60 gallons' capacity. Originally two churns were used, but only one is shown. It is on record that the churn was at work in 1848, being used twice a week and churning 46 dozen pounds of butter per week. There is evidence that the churn was working as late as 1879. It is stated that the rate of working of the churn was 45 revolutions per minute, which would necessitate the good speed of 2 to 2·25 miles per hour on the part of the horse, and that the time of churning was from 1·5 to 2 hours.

The whole apparatus is an excellent specimen of the kind of work done by millwrights in the late eighteenth and early nineteenth centuries. Inv. 1927–1090.

213. CHURN. Lent by The Disc Churn Co.

This is a one-gallon churn, fitted with glass to show the internal construction of the machine, which is known as a disc churn.

It consists of a box, with a cover having two internal flaps, and containing a sharpedged wooden disc capable of being rapidly rotated by hand through external spur gearing having a ratio of 6 to 1.

The motion of the disc, and the continual projection of the milk against the cover and casing, break up the envelopes of the fat cells in the cream and liberate and collect the butter. The butter-milk is run off by an outlet at the bottom of the churn.

Inv. 1894–157.

214. MECHANICAL PLUNGER CHURNS. (Scale 1 : 4.) Lent by The Royal Scottish Museum, Edinburgh, and The University of Edinburgh.

These three models were submitted to the Highland Society early in the nineteenth century, and were in their possession in 1832. In each case the actual churning is carried out by means of paired plungers reciprocating vertically in a pair of churns.

Inv. 1928–368, 513, 641.

Horse-driven Butter Churn (Cat. No. 212).

Horse-driven Butter Churn (Cat. No. 212).

215. MECHANICAL PADDLE CHURN. (Scale 1 : 4.) Lent by The Royal Scottish Museum.

This example of a horse-driven paddle churn was introduced later than the mechanical plungers, but was in existence before 1857.

The horse pulls by means of the wooden fork and rotates the vertical beam in the middle of the machine. The large horizontal gear wheel mounted on the beam drives the two small gear wheels which drive the paddles inside the churns. Inv. 1928–367.

216. CHURN. Lent by George Hathaway, Esq.

This is the smallest, or No. 1, size of the Shakespearian end-over-end eccentric-motion churn.

The churn is first well washed with boiling water and soda, and then with pure water, cold in warm weather and hot in cold weather. The cream should be at a temperature of from 50 to 60 deg. F. and the rate of revolution should be 50 revs. per min. When the butter comes, the churn is turned backward and forward for a few minutes until the butter is collected, the milk is then drawn off, the butter is well washed with cold water, and is lifted out of the churn in a granular state. The churn is then immediately cleaned with cold water and scalded.

The capacity of this churn is 5 gal., and the amount of cream which it will churn at once is 0·25 to 2 gal., producing 0·5 to 6 lb. of butter. Inv. 1914–1.

217. BUTTER MOULDS. Lent by The Royal Scottish Museum.

This is a collection of various utensils used in preparing butter for the table.

As a rule, for convenience of packing, butter is made up into brick-shaped pats and a criss-cross pattern is worked on the upper surface with the edge of a Scotch hand. Sometimes the circular pat is made, when a special mould is used, and a design is printed by the mould on the upper surface. The moulds in this collection are intended for putting up butter in fancy patterns, among them are such designs as, the Coats of Arms of Scotland and of the Duke of Edinburgh, a large swan mould and a small one, a mould for making butter balls and other moulds of heraldic designs. Inv. 1928–464–474, 477–483.

218. CHEESE MOULD. Lent by The Royal Scottish Museum.

This is one kind of mould which is used to contain the curd when it is being pressed.

The curd is packed into the mould and the vessel is placed on the press. As the pressure applied to hard cheeses may be in the region of three tons, it is necessary that the mould should be stoutly built. The more modern mould is usually made of metal. Inv. 1928–421.

219. CHESSART. Lent by The Royal Scottish Museum.

This type of wooden bowl is still used in the process of making soft cheeses.

The pressure required, particularly when the work is done on the small scale, is not sufficient to warrant the use of a press. The whey is freed from the curd by placing it in the bowl and pressing down on it with the wooden disc. Inv. 1928–422.

220. CURD BREAKER. (Scale 1 : 4.) Lent by The Royal Scottish Museum.

This type of machine was invented between 1830 and 1840 for breaking up the curd into small pieces in the process of making skim-milk cheeses.

Its essential parts consist of a wooden cylinder which is studded with iron spikes, and which can be turned by the handle, and of two pieces of hardwood which fill the spaces between the roller and the sides of the hopper, and which are notched to permit the passage of the spikes of the roller. The hopper is filled with curd roughly broken by hand, and the operator turns the handle with one hand and presses the curd down with the other. The machine is made so that it can be easily dismantled for cleaning purposes. Inv. 1928–521.

221. CHEESE PRESS.

This is an example of an eighteenth century cheese press, which obtained its pressure by direct means.

The curd was packed into a container, probably of wood and open at both ends, and was placed on the lower horizontal piece of timber. A strong piece of wood

89

fitting not too tightly in the upper end of the container, is inserted, its lower end resting on the cloths above the cheese, and the upper end protruding from the container.

The upper horizontal piece of timber is lowered and allowed to rest on the piece of wood, thus applying the pressure of the weight of the timber to the cheese. Further pressure is applied by placing blocks of stone, weights, etc., on the upper platform.

Inv. 1927–986.

222. CHEESE PRESS. (Replica copied from an original in the Museum, Torquay.)

This type of press, introduced by Brown, probably in the late eighteenth century, shows a mechanical means of applying pressure to the cheese.

In this case a vertical ratchet is used instead of a screw, and by turning the small handle the plunger can be lowered on to the cheese. When the latter is in position the lever is raised to an angle of 45 deg. above the horizontal, and is held in that position by allowing the pawl to engage with the toothed wheel. A weight hung on the end of the lever will react on the curd and the pressure will also be maintained until the lever has moved through a right angle. If it is required to maintain the pressure further the lever is raised, the pawl re-engaged and the weight re-applied.

Inv. 1927–999.

223. CHEESE PRESS. (Scale 1 : 4.) Lent by The Royal Scottish Museum.

This model represents a type of press which was made and used probably early in the nineteenth century.

Pressure on the curd is applied by means of a rather complicated system of levers and is maintained by hanging a weight on the long lever which projects at the side of the press.

Inv. 1928–557.

224. CHEESE PRESS. (Scale 1 : 4.) Lent by The Royal Scottish Museum.

This is a model of a cheese press which was introduced into Scotland in 1833. The curd is placed in a mould on the bottom plate and the top plate is brought down to apply pressure to it.

The pressure is applied by turning the winch handle ; this turns a small gear wheel which drives the larger one on the second shaft. Rotation of the second shaft brings into action the pinion and rack which move the top plate up or down. The top plate is lowered by turning the handle until the resistance of the cheese becomes too great, when further pressure is applied by raising and depressing the lever which carries the weight. This lever when depressed drives by means of a pall the ratchet wheel which is mounted on the first shaft, and so makes it possible to lower the plate after the resistance is too great to be overcome by turning the handle.

The pressure is maintained, as the mass of curd shrinks, by the action of the weight on the lever.

Inv. 1928–462.

225. CHEESE PRESS. (Scale 1 : 4.) Lent by The Royal Scottish Museum.

In the type of press represented by this model an attempt has been made to make use of the principle of the screw, a principle which is found embodied in presses by the middle of the nineteenth century. It is, therefore, probable that this press was introduced about that time.

The method employed would allow of lighter construction of the press than would be the case with earlier types, and the work of the dairymaid in applying the pressure would be a great deal easier. The idea, however, required development as there was no means of maintaining the pressure on the shrinking bulk of the cheese.

Inv. 1928–420

226. CHEESE RACK. (Scale 1 : 4.) Lent by The Royal Scottish Museum.

This is a model of a rack, invented in the middle of the nineteenth century, in which cheeses are stored, and which is constructed so that they can be turned automatically.

The final stage in the process of making hard cheese such as Cheddar consists of storing the cheeses for some time in a cool place, usually a cellar. During this period the cheeses are maturing, and it is necessary to turn them upside-down daily so that the upper end does not become too dry and the cheeses mature unevenly. The common practice is to keep the cheeses on wooden shelves and to turn each one by hand, but this rack is intended to expedite the process of turning. The shelves are made so that standard size cheeses fit between them easily and air can circulate round them freely. The whole rack pivots on its centre, it is held by a catch in the upright position, and the process of turning, consists of letting go the catch, rotating the rack through 180 degrees and fixing it again in an upright position.   Inv. 1928–369.

227. CHEESE PRESS. (Scale 1 : 4.) Lent by The Royal Scottish Museum.

This type of press was evolved in the latter half of the nineteenth century.

The pressure is applied by means of a rack and pinion, the former being carried on the vertical shaft which supports the plunger. The short horizontal shaft which carries the pinion also carries a ratchet which is turned by a pawl and lever. This lever is connected to a longer lever, which when the plunger is to be lowered is worked by hand. The pressure is maintained by hanging a weight on the long hand lever.

Inv. 1928–373.

BIBLIOGRAPHY

Books and Journals consulted in connection with this handbook.

BOOKS—
BACON. Report on the Agriculture of Norfolk, 1884.
BLITH, WALTER. The English Improver Improved, 1653.
BOITARDS. Les Instrumens Aratoires, 1883.
BOND, J. R. Farm Implements and Machinery, 1923.
BROWN, ROBERT. The Compleat Farmer, 1807.
CASSON, H. N. Cyrus Hall McCormick, 1909.
COMMON, R. J. F. The History of the Invention of the Reaper, 1907.
DONALDSON, J. Agricultural Biography.
DUHAMEL DU MONCEAU. The Elements of Agriculture, 1764.
FITZHERBERT, JOHN. The Book of Husbandry, 1523 (c).
GOOGE, BARNABY. The First Book of Husbandry, 1600.
HALE, THOMAS. The Compleat Body of Husbandry, 1756.
HARTLIB. The Husbandry of Brabant and Flanders.
HOOPS. Waldbaume und Kulturpflanzen im Germanischem Altertum.
LOUDON, J. C. Encyclopedia of Agriculture, 1831.
MARSHALL. The Rural Economy of Norfolk.
MARSHALL. The Rural Economy of the West of England, 1798.
PETERS, MATTHEW. The Rational Farmer.
RANSOME, J. E. Ploughs and Ploughing, 1863.
RANSOME, J. E. Double-Furrow Ploughs, 1872.
SCOTT. A Text-Book of Farm Engineering, 1885.
SKAPPEL, SIMEN. Observations on the History of Norwegian Agriculture.
SMALL, JAMES. The Plough, 1784.
SOMERVILLE. Sheep, Ploughs and Oxen, 1809.
STEPHENS, H. The Book of the Farm, 1855.
TULL, JETHRO. The Book of Husbandry, 1721.
WOODCROFT, B. Reaping Machines, 1853.
YOUNG, ARTHUR. Eastern Tour, 1771.
YOUNG, ARTHUR. Annals of Agriculture.

JOURNALS—
Bath and West of England Agricultural Society.
Highland and Agricultural Society of Scotland.
Highland Society of Scotland, Prize Essays and Transactions.
Institution of Mechanical Engineers.
New York State Agricultural Society.
Paris Exposition Universelle, de 1900.
Royal Agricultural Society of England.
Royal Society of Arts.

PERIODICALS—
Agricultural Implement and Machinery Review.

PATENTS—
Abridgments and Specifications of British Patents.
U.S.A. Patent Specifications.

INDEX

93

Printed in Great Britain
by Amazon

32816412R00076